SOIL MECHANICS LABORATORY MANUAL

Sixth Edition

Braja M. Das
Dean, College of Engineering and Computer Science
California State University, Sacramento

New York Oxford
OXFORD UNIVERSITY PRESS
2002

Oxford University Press

Oxford New York
Athens Auckland Bangkok Bogotá Buenos Aires Calcutta
Cape Town Chennai Dar es Salaam Delhi Florence Hong Kong Istanbul
Karachi Kuala Lumpur Madrid Melbourne Mexico City Mumbai
Nairobi Paris São Paulo Shanghai Singapore Taipei Tokyo Toronto Warsaw

and associated companies in
Berlin Ibadan

Published by Oxford University Press, Inc.
198 Madison Avenue, New York, New York 10016
http://www.oup-usa.org

ISBN: 0-19-515046-5

Printing (last digit): 9 8 7 6 5 4 3 2

Printed in the United States of America
on acid-free paper

To Janice and Valerie

CONTENTS

1. Laboratory Test and Report Preparation **1**
2. Determination of Water Content **5**
3. Specific Gravity **9**
4. Sieve Analysis **15**
5. Hydrometer Analysis **23**
6. Liquid Limit Test **35**
7. Plastic Limit Test **41**
8. Shrinkage Limit Test **45**
9. Engineering Classification of Soils **51**
10. Constant Head Permeability Test in Sand **69**
11. Falling Head Permeability Test in Sand **75**
12. Standard Proctor Compaction Test **81**
13. Modified Proctor Compaction Test **89**
14. Determination of Field Unit Weight of Compaction by Sand Cone Method **93**
15. Direct Shear Test on Sand **99**
16. Unconfined Compression Test **109**
17. Consolidation Test **117**
18. Triaxial Tests in Clay **129**

References **145**

Appendices

A. Weight-Volume Relationships **147**
B. Data Sheets for Laboratory Experiments **151**
C. Data Sheets for Preparation of Laboratory Reports **215**

PREFACE

Since the early 1940's the study of soil mechanics has made great progress all over the world. A course in soil mechanics is presently required for undergraduate students in most four-year civil engineering and civil engineering technology programs. It usually includes some laboratory procedures that are essential in understanding the properties of soils and their behavior under stress and strain; the present laboratory manual is prepared for classroom use by undergraduate students taking such a course.

The procedures and equipment described in this manual are fairly common. For a few tests such as permeability, direct shear, and unconfined compression, the existing equipment in a given laboratory may differ slightly. In those cases, it is necessary that the instructor familiarize students with the operation of the equipment. Triaxial test assemblies are costly, and the equipment varies widely. For that reason, only general outlines for triaxial tests are presented.

For each laboratory test procedure described, sample calculation(s) and graph(s) are included. Also, blank tables for each test are provided at the end of the manual for student use in the laboratory and in preparing the final report. The accompanying diskette contains the Soil Mechanics Laboratory Test Software, a stand-alone program that students can use to collect and evaluate the data for each of the 18 labs presented in the book. For this new edition, Microsoft Excel templates have also been provided for those students who prefer working with this popular spreadsheet program.

Professor William Neuman of the Department of Civil Engineering at California State University, Sacramento, took most of the photographs used in this edition. Thanks are due to Professor Cyrus Aryani of the Department of Civil Engineering at California State University, Sacramento, for his assistance in taking the photographs. Last, I would like to thank my wife, Janice F. Das, who apparently possesses endless energy and enthusiasm. Not only did she type the manuscript, she also prepared all of the tables, graphs, and other line drawings.

Braja M. Das
dasb@csus.edu

1 Laboratory Test and Preparation of Report

Introduction

Proper laboratory testing of soils to determine their physical properties is an integral part in the design and construction of structural foundations, the placement and improvement of soil properties, and the specification and quality control of soil compaction works. It needs to be kept in mind that natural soil deposits often exhibit a high degree of nonhomogenity. The physical properties of a soil deposit can change to a great extent even within a few hundred feet. The fundamental theoretical and empirical equations that are developed in soil mechanics can be properly used in practice if, and only if, the physical parameters used in those equations are properly evaluated in the laboratory. So, learning to perform laboratory tests of soils plays an important role in the geotechnical engineering profession.

Use of Equipment

Laboratory equipment is never cheap, but the cost may vary widely. For accurate experimental results, the equipment should be properly maintained. The calibration of certain equipment, such as balances and proving rings, should be checked from time to time. It is always necessary to see that all equipment is clean both before and after use. Better results will be obtained when the equipment being used is clean, so always maintain the equipment as if it were your own.

Recording the Data

In any experiment, it is always a good habit to record all data in the proper table immediately after it has been taken. Oftentimes, scribbles on scratch paper may later be illegible or even misplaced, which may result in having to conduct the experiment over, or in obtaining inaccurate results.

Report Preparation

In the classroom laboratory, most experiments described herein will probably be conducted in small groups. However, the laboratory report should be written by each student individually. This is one way for students to improve their technical writing skills. Each report should contain:

1. Cover page—This page should include the title of the experiment, name, and date on which the experiment was performed.
2. Following the cover page, the items listed below should be included in the body of the report:
   a. Purpose of the experiment
   b. Equipment used
   c. A schematic diagram of the main equipment used
   d. A brief description of the test procedure
3. Results—This should include the data sheet(s), sample calculations(s), and the required graph(s).
4. Conclusion—A discussion of the accuracy of the test procedure should be included in the conclusion, along with any possible sources of error.

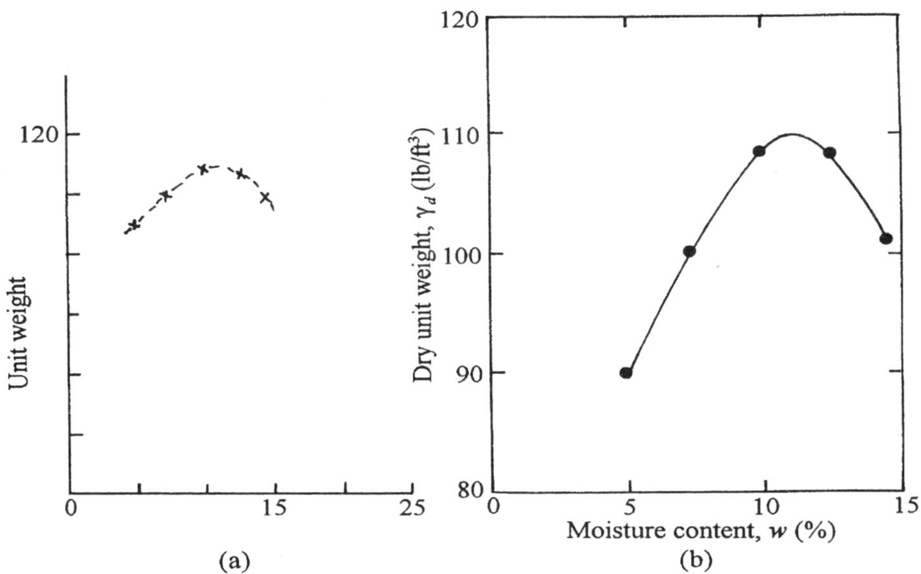

Figure 1-1.
(a) A poorly drawn graph for dry unit weight of soil vs. moisture content
(b) The results given in (a), drawn in a more presentable manner

Graphs and Tables Prepared for the Report

Graphs and tables should be prepared as neatly as possible. *Always* give the units. Graphs should be made as large as possible, and they should be properly labeled. Examples of a poorly-drawn graph and an acceptable graph are shown in Fig. 1–1. When necessary, French curves and a straight edge should be used in preparing graphs.

Table 1–1. Conversion Factors

	English to SI Units		SI to English Units	
Quantity	English	SI	SI	English
Length	1 in. 1 ft	25.4 mm 0.3048 m 304.8 mm	1 mm 1 m	3.937×10^{-2} in. 3.281×10^{-3} ft 39.37 in. 3.281 ft
Area	1 in.^{2} 1 ft^2	6.4516×10^{-4}m^2 6.4516 cm^2 645.16 mm^2 929×10^{-4}m^2 929.03 cm^2 92903 mm^2	1 cm^2 1 m^2	0.155 in.^{2} 1.076×10^{-3} ft^2 1550 in.^{2} 10.76 ft^2
Volume	1 in.^{3} 1 ft^3 1 ft^3	16.387 cm^3 0.028317 m^3 28.3168 l	1 cm^3 1 m^3	0.061 in.^{3} 3.531×10^{-5} ft^3 61023.74 in.^{3} 35.315 ft^3
Velocity	1 ft/s 1 ft/min	304.8 mm/s 0.3048 m/s 5.08 mm/s 0.00508 m/s	1 cm/s	1.969 ft/min 1034643.6 ft/year
Force	1 lb	4.448 N	1 N 1 kN	0.22482 lb 0.22482 kip
Stress	1 lb/in.^{2} 1 lb/ft^2	6.9 kN/m^2 47.88 N/m^2	1 kN/m^2	0.145 lb/in.^{2} 2.089×10^{-2} lb/ft^2
Unit Weight	1 lb/ft^3	157.06 N/m^3	1 kN/m^3	6.367 lb/ft^3
Coefficient of Consolidation	1 in.^{2}/s 1 ft^2/s	6.452 cm^2/s 929.03 cm^2/s	1 cm^2/s	0.155 in.^{2}/s 2.883×10^3 ft^2/month
Mass			1 kg	2.2046 lb 2.2046×10^{-3} kip

Units

It may be necessary to express the results of laboratory tests in a given system of units. At this time in the United States, both the English and the SI system of units are used. Conversion of units may be necessary in preparing reports. Some selected conversion factors from the English to the SI units and from SI to English units are given in Table 1–1.

Standard Test Procedures

In the United States, most laboratories conducting tests on soils for engineering purposes follow the procedures outline by the American Society for Testing and Materials (ASTM) and the American Association of State Highway and Transportation Officials (AASHTO). The procedures and equipment for soil tests may vary slightly from laboratory to laboratory, but the basic concepts remain the same. The test procedures described in this manual may not be exactly the same as specified by ASTM or AASHTO; however, for the students, it is beneficial to know the standard test designations and compare them with the laboratory work actually done. For this reason some selected AASHTO and ASTM standard test designations are given in Table 1–2.

Table 1–2. Some Important AASHTO and ASTM Soil Test Designations

Name of Test	AASHTO Test Designation	ASTM Test Designation
Water content	T-265	D-2216
Specific gravity	T-100	D-854
Sieve analysis	T-87, T-88	D-421
Hydrometer analysis	T-87, T-88	D-422
Liquid limit	T-89	D-4318
Plastic limit	T-90	D-4318
Shrinkage limit	T-92	D-427
Standard Proctor compaction	T-99	D-698
Modified Proctor compaction	T-180	D-1557
Field density by sand cone	T-191	D-1556
Permeability of granular soil	T-215	D-2434
Consolidation	T-216	D-2435
Direct shear (granular soil)	T-236	D-3080
Unconfined compression	T-208	D-2166
Triaxial	T-234	D-2850
AASHTO Soil Classification System	M-145	D-3282
Unified Soil Classification System	—	D-2487

2
Determination of Water Content

Introduction

Most laboratory tests in soil mechanics require the determination of water content. Water content is defined as

$$w = \frac{\text{weight of water present in a given soil mass}}{\text{weight of dry soil}} \tag{2.1}$$

Water content is usually expressed in percent.

For better results, the *minimum* size of the most soil specimens should be approximately as given in Table 2–1. These values are consistent with ASTM Test Designation D-2216.

Table 2–1. Minimum Size of Moist Soil Samples to Determine Water Content

Maximum Particle Size in the Soil (mm)	U.S. Sieve No.	Minimum Mass of Soil Sample (g)
0.425	40	20
2.0	10	50
4.75	4	100
9.5	3/8 in.	500
19.0	3/4 in.	2500

Equipment

1. Moisture can(s).

   Moisture cans are available in various sizes [for example, 2-in. (50,8 mm) diameter and ⅞ in. (22.2 mm) high, 3.5-in. (88.9 mm) diameter and 2 in. (50.8 mm) high].
2. Oven with temperature control.

   For drying, the temperature of oven is generally kept between 105°C to 110°C. A higher temperature should be avoided to prevent the burning of organic matter in the soil.
3. Balance.

   The balance should have a readability of 0.01 g for specimens having a mass of 200 g or less. If the specimen has a mass of over 200 g, the readability should be 0.1 g.

Procedure

1. Determine the mass (g) of the empty moisture can plus its cap (W_1), and also record the number.
2. Place a sample of representative moist soil in the can. Close the can with its cap to avoid loss of moisture.
3. Determine the combined mass (g) of the closed can and moist soil (W_2).
4. Remove the cap from the top of the can and place it on the bottom (of the can).
5. Put the can (Step 4) in the oven to dry the soil to a constant weight. In most cases, 24 hours of drying is enough.
6. Determine the combined mass (g) of the dry soil sample plus the can and its cap (W_3).

Calculation

1. Calculate the mass of moisture = $W_2 - W_3$
2. Calculate the mass of dry soil = $W_3 - W_1$
3. Calculate the water content

$$w\ (\%) = \frac{W_2 - W_3}{W_3 - W_1} \times 100 \tag{2.2}$$

Report the water content to the nearest 1% or 0.1% as appropriate based on the size of the specimen.

A sample calculation of water content is given in Table 2–2.

Table 2–2. Determination of Water Content

Description of soil *Brown silty clay* Sample No. *4*

Location

Tested by Date

Item	Test No.		
	1	2	3
Can No.	42	31	54
Mass of can, W_1 (g)	17.31	18.92	16.07
Mass of can + wet soil, W_2 (g)	43.52	52.19	39.43
Mass of can + dry soil, W_3 (g)	39.86	47.61	36.13
Mass of moisture, $W_2 - W_3$ (g)	3.66	4.58	3.30
Mass of dry soil, $W_3 - W_1$ (g)	22.55	28.69	20.06
Moisture content, $w(\%) = \dfrac{W_2 - W_3}{W_3 - W_1} \times 100$	16.2	16.0	16.5

Average moisture content, w *16.2* %

General Comments

a. Most natural soils, which are sandy and gravelly in nature, may have water contents up to about 15 to 20%. In natural fine-grained (silty or clayey) soils, water contents up to about 50 to 80% can be found. However, peat and highly organic soils with water contents up to about 500% are not uncommon.

Typical values of water content for various types of natural soils in a saturated state are shown in Table 2–3.

b. Some organic soils may decompose during oven drying at 110°C. An oven drying temperature of 110° may be too high for soils containing gypsum, as this material slowly dehydrates. According to ASTM, a drying temperature of 60°C is more appropriate for such soils.

c. Cooling the dry soil after oven drying (Step 5) in a desiccator is recommended. It prevents absorption of moisture from the atmosphere.

Table 2–3. Typical Values of Water Content in a Saturated State

Soil	Natural Water Content in a Saturated State (%)
Loose uniform sand	25–30
Dense uniform sand	12–16
Loose angular-grained silty sand	25
Dense angular-grained silty sand	15
Stiff clay	20
Soft clay	30–50
Soft organic clay	80–130
Glacial till	10

3
Specific Gravity of Soil Solids

Introduction

The specific gravity of a given material is defined as the ratio of the weight of a given volume of the material to the weight of an equal volume of distilled water. In soil mechanics, the specific gravity of soil solids (which is often referred to as the specific gravity of soil) is an important parameter for calculation of the weight–volume relationship. Thus specific gravity, G_s, is defined as

$$G_s = \frac{\text{unit weight (or density) of soil solids only}}{\text{unit weight (or density) or water}}$$

or

$$G_s = \frac{W_s / V_s}{\rho_2} = \frac{W_s}{V_s \rho_w} \tag{3.1}$$

where W_s = mass of soil solids (g)
V_s = volume of soil solids (cm^3)
ρ_w = density of water (g/cm^3).

The general ranges of the values of G_s for various soils are given in Table 3–1. The procedure for determination of specific gravity, G_s, described here is applicable for soils composed of *particles smaller than 4.75 mm* (No. 4 U.S. sieve) *in size*.

Table 3–1. General Ranges of G_s for Various Soils

Soil Type	Range of G_s
Sand	2.63–2.67
Silts	2.65–2.7
Clay and silty clay	2.67–2.9
Organic soil	less than 2

Equipment

1. Volumetric flask (500 ml)
2. Thermometer graduated in 0.5°C division scale
3. Balance sensitive up to 0.01 g
4. Distilled water
5. Bunsen burner and a stand (and/or vacuum pump or aspirator)
6. Evaporating dishes
7. Spatula
8. Plastic squeeze bottle
9. Drying oven

The equipment for this experiment is shown in Fig. 3–1.

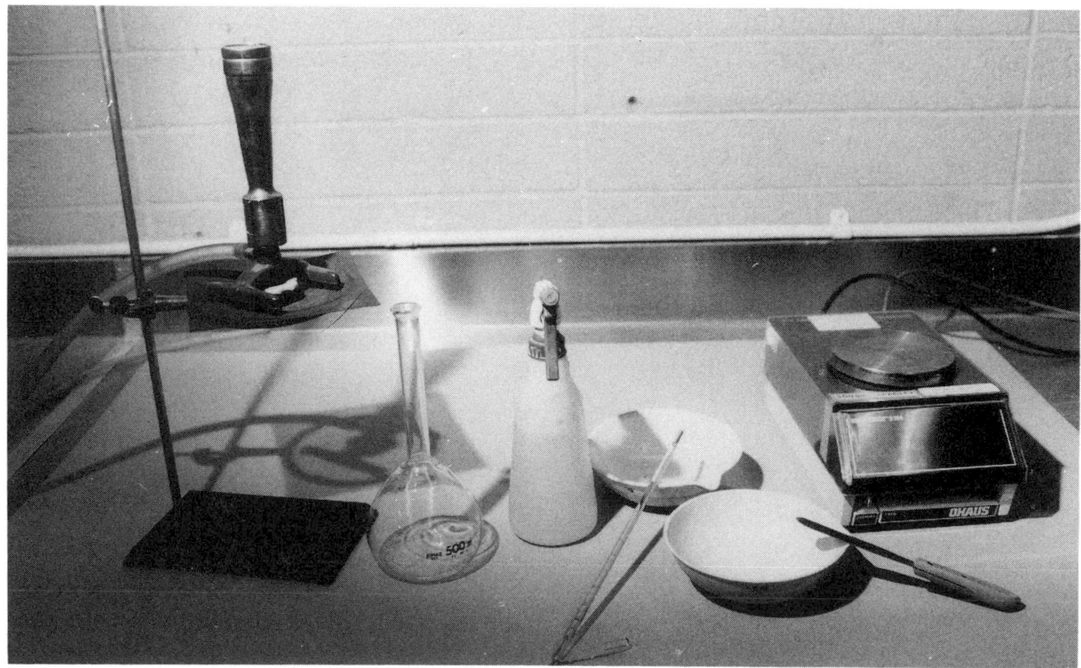

Figure 3–1. Equipment for conducting specific gravity test.

Procedure

1. Clean the volumetric flask well and dry it.
2. Carefully fill the flask with de-aired, distilled water up to the 500 ml mark (bottom of the meniscus should be at the 500 ml mark).
3. Determine the mass of the flask and the water filled to the 500 ml mark (W_1).
4. Insert the thermometer into the flask with the water and determine the temperature of the water $T = T_1$°C.
5. Put *approximately* 100 grams of air dry soil into an evaporating dish.
6. If the soil is cohesive, add water (de-aired and distilled) to the soil and mix it to the form of a smooth paste. Keep it soaked for about one-half to one hour in the evaporating dish. (*Note:* This step is not necessary for granular, i.e., noncohesive, soils.)
7. Transfer the soil (if granular) or the soil paste (if cohesive) into the volumetric flask.
8. Add distilled water to the volumetric flask containing the soil (or the soil paste) to make it about two-thirds full.
9. Remove the air from the soil-water mixture. This can be done by:
    a. Gently boiling the flask containing the soil-water mixture for about 15 to 20 minutes. Accompany the boiling with continuous agitation of the flask. (If too much heat is applied, the soil may boil over.) Or
    b. Apply vacuum by a vacuum pump or aspirator until all of the entrapped air is out.

    This is an *extremely important step.* Most of the errors in the results of this test are due to *entrapped air which is not removed.*
10. Bring the temperature of the soil-water mixture in the volumetric flask down to room temperature, i.e., T_1°C—see Step 4. (This temperature of the water is at room temperature.)
11. Add de-aired, distilled water to the volumetric flask until the bottom of the meniscus touches the 500 ml mark. Also dry the outside of the flask and the inside of the neck above the meniscus.
12. Determine the combined mass of the bottle plus soil plus water (W_2).
13. Just as a precaution, check the temperature of the soil and water in the flask to see if it is $T_1° \pm 1$°C or not.
14. Pour the soil and water into an evaporating dish. Use a plastic squeeze bottle and wash the inside of the flask. Make sure that no soil is left inside.
15. Put the evaporating dish in a oven to dry to a constant weight.
16. Determine the mass of the dry soil in the evaporating dish (W_s).

Calculation

1. Calculate the specific gravity

$$G_s = \frac{\text{mass of soil, } W_s}{\text{mass of equal volume of soil}}$$

where

$$\text{mass of soil} = W_s$$

$$\text{mass of equal volume of water, } W_w = (W_1 + W_s) - W_2$$

So

$$G_{s(\text{at } T_1{}^\circ\text{C})} = \frac{W_s}{W_w} \tag{3.2}$$

Specific gravity is generally reported on the value of the density of water at 20°C. So

$$G_{s(\text{at } 20^\circ\text{c})} = G_{s(\text{at } T_1{}^\circ\text{C})} \left[\frac{\rho_{w(\text{at } T_1{}^\circ\text{C})}}{\rho_{w(\text{at } 20^\circ\text{C})}} \right] = G_{s(\text{at } T_1{}^\circ\text{C})} A \tag{3.3}$$

$$\text{where } A = \frac{\rho_{w\,(\text{at } T_1{}^\circ\text{C})}}{\rho_{w\,(\text{at } T_{20}{}^\circ\text{C})}} \tag{3.4}$$

ρ_w = density of water.

The values of A are given in Table 3–2.

Table 3–2. Values of *A* [Eq. (3.4)]

Temperature (T_1°C)	*A*	Temperature (T_1°C)	*A*
16	1.0007	24	0.9991
17	1.0006	25	0.9988
18	1.0004	26	0.9986
19	1.0002	27	0.9983
20	1.0000	28	0.9980
21	0.9998	29	0.9977
22	0.9996	30	0.9974
23	0.9993		

At least three specific gravity tests should be conducted. For correct results, these values should not vary by more than 2 to 3%. A sample calculation for specific gravity is shown in Table 3–3.

Table 3–3. Specific Gravity of Soil Solids

Description of soil *Light brown sandy silt* Sample No. *23*

Volume of flask at 20°C *500* ml Temperature of test *23* °C *A* *0.9993* (Table 3–2)

Location________________

Tested by ________________ Date________________

Item	Test No.		
	1	2	3
Volumetric flask No.	*6*	*8*	*9*
Mass of flask + water filled to mark, W_1 (g)	*666.0*	*674.0*	*652.0*
Mass of flask + soil + water filled to mark, W_2 (g)	*722.0*	*738.3*	*709.93*
Mass of dry soil, W_s (g)	*99.0*	*103.0*	*92.0*
Mass of equal volume of water as the soil solids, W_w (g) = $(W_1 + W_s) - W_2$	*37.0*	*38.7*	*34.07*
$G_{s(T_1°C)} = W_s / W_w$	*2.68*	*2.66*	*2.70*
$G_{s(20°C)} = G_{s(T_1°C)} \times A$	*2.68*	*2.66*	*2.70*

Average G_s $\frac{(2.68 + 2.66 + 2.70)}{3} = 2.68$

4

Sieve Analysis

Introduction

In order to classify a soil for engineering purposes, one needs to know the distribution of the size of grains in a given soil mass. Sieve analysis is a method used to deter mine the grain-size distribution of soils. Sieves are made of woven wires with square openings. Note that as the sieve number increases the size of the openings decreases. Table 4–1 gives a list of the U.S. standard sieve numbers with their corresponding size of openings. For all practical purposes, the No. 200 sieve is the sieve with the smallest opening that should be used for the test. The sieves that are most commonly used for soil tests have a diameter of 8 in. (203 mm). A stack of sieves is shown in Fig. 4.–1.

The method of sieve analysis described here is applicable for soils that are *mostly granular with some or no fines*. Sieve analysis does not provide information as to shape of particles.

Table 4–1. U.S. Sieve Sizes

Sieve No.	Opening (mm)	Sieve No.	Opening (mm)
4	4.75	35	0.500
5	4.00	40	0.425
6	3.35	45	0.355
7	2.80	50	0.300
8	2.36	60	0.250
10	2.00	70	0.212
12	1.70	80	0.180
14	1.40	100	0.150
16	1.18	120	0.125
18	1.00	140	0.106
20	0.85	200	0.075
25	0.71	270	0.053
30	0.60	400	0.038

Figure 4–1. A stack of sieves with a pan at the bottom and a cover on the top.

Equipment

1. Sieves, a bottom pan, and a cover
   Note: Sieve numbers 4, 10, 20, 40, 60, 140, and 200 are generally used for most standard sieve analysis work.
2. A balance sensitive up to 0.1 g
3. Mortar and rubber-tipped pestle
4. Oven
5. Mechanical sieve shaker

Procedure

1. Collect a *representative* oven dry soil sample. Samples having largest particles of the size of No. 4 sieve openings (4.75 mm) should be about 500 grams. For soils having largest particles of size greater than 4.75 mm, larger weights are needed.
2. Break the soil sample into individual particles using a mortar and a rubber-tipped pestle. (*Note:* The idea is to break up the soil into individual particles, not to break the particles themselves.)
3. Determine the mass of the sample accurately to 0.1 g (W).

Figure 4–2. Washing of the soil retained on No. 200 sieve.

4. Prepare a stack of sieves. A sieve with larger openings is placed above a sieve with smaller openings. The sieve at the bottom should be No. 200. A bottom pan should be placed under sieve No. 200. As mentioned before, the sieves that are generally used in a stack are Nos. 4, 10, 20, 40, 60, 140, and 200; however, more sieves can be placed in between.
5. Pour the soil prepared in Step 2 into the stack of sieves from the top.
6. Place the cover on the top of the stack of sieves.
7. Run the stack of sieves through a sieve shaker for about 10 to 15 minutes.
8. Stop the sieve shaker and remove the stack of sieves.
9. Weigh the amount of soil retained on each sieve and the bottom pan.
10. If a *considerable* amount of soil with silty and clayey fractions is retained on the No. 200 sieve, it has to be washed. Washing is done by taking the No. 200 sieve with the soil retained on it and pouring water through the sieve from a tap in the laboratory (Fig. 4–2).

When the water passing through the sieve is clean, stop the flow of water. Transfer the soil retained on the sieve at the end of washing to a porcelain evaporating dish by back washing (Fig. 4–3). Put it in the oven to dry to a constant weight. (*Note:* This step is not necessary if the amount of soil retained on the No. 200 sieve is small.)

Determine the mass of the dry soil retained on No. 200 sieve. The difference between this mass and that retained on No. 200 sieve determined in Step 9 is the mass of soil that has washed through.

Figure 4–3. Back washing to transfer the soil retained on No. 200 sieve to an evaporating dish.

Calculation

1. Calculate the percent of soil retained on the nth sieve (counting from the top)

$$= \frac{\text{mass retained, } W_n}{\text{total mass, } W \text{ (Step 3)}} \times 100 = R_n \tag{4.1}$$

2. Calculate the cumulative percent of soil retained on the nth sieve

$$= \sum_{i=1}^{i=n} R_n \tag{4.2}$$

3. Calculate the cumulative percent passing through the nth sieve

$$= \text{percent finer} = 100 - \sum_{i=1}^{i=n} R_n \tag{4.3}$$

Note: If soil retained on No.200 sieve is washed, the dry unit weight determined after washing (Step 10) should be used to calculate percent finer (than No. 200 sieve). The weight lost due to washing should be added to the weight of the soil retained on the pan.

A sample calculation of sieve analysis is shown in Table 4–2.

Table 4–2. Sieve Analysis

Description of soil *Sand with some fines* Sample No. *2*

Mass of oven dry specimen, W *500* g

Location______

Tested by______ Date______

Sieve No.	Sieve opening (mm)	Mass of soil retained on each sieve, W_n (g)	Percent of mass retained on each sieve, R_n	Cumulative percent retained, $\sum R_n$	Percent finer, $100-\sum R_n$
4	4.750	0	0	0	100.0
10	2.000	40.2	8.0	8.0	92.0
20	0.850	84.6	16.9	24.9	75.1
30	0.600	50.2	10.0	34.9	65.1
40	0.425	40.0	8.0	42.9	57.1
60	0.250	106.4	21.3	64.2	35.8
140	0.106	108.8	21.8	86.0	14.0
200	0.075	59.4	11.9	97.9	2.1
Pan	——	8.7			

$\sum$ *498.3* $= W_1$

Mass loss during sieve analysis = $\dfrac{W - W_1}{W} \times 100$ = *0.34* % (OK if less than 2%)

Graphs

The grain-size distribution obtained from the sieve analysis is plotted in a semi-logarithmic graph paper with grain size plotted on the log scale and percent finer plotted on the natural scale. Figure 4–4 is a grain-size distribution plot for the calculation shown in Table 4–2.

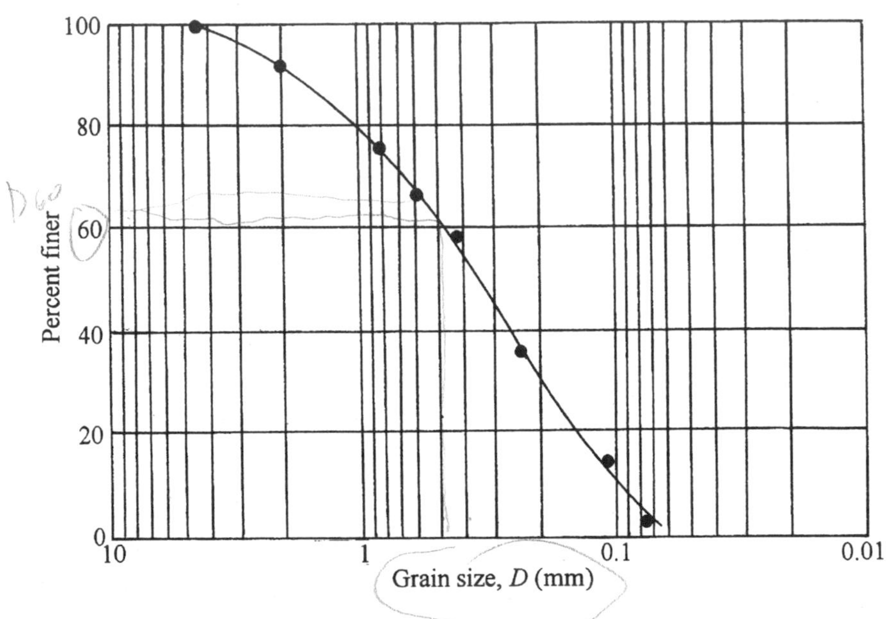

Figure 4–4. Plot of percent finer vs. grain size from the calculation shown in Table 4–2.

The grain-size distribution plot helps to estimate the percent finer than a given sieve size which might not have been used during the test.

Other Calculations

1. Determine D_{10}, D_{30}, and D_{60} (from Fig. 4–4), which are, respectively, the diameters corresponding to percents finer of 10%, 30%, and 60%.
2. Calculate the uniformity coefficient (C_u) and the coefficient of gradation (C_c) using the following equations:

$$C_u = \frac{D_{60}}{D_{10}} \tag{4.4}$$

$$C_c = \frac{D_{30}^2}{D_{60} \times D_{10}} \tag{4.5}$$

As an example, from Fig. 4–4, $D_{60} = 0.46$ mm, $D_{30} = 0.21$ mm, and $D_{10} = 0.098$ mm. So

$$C_u = \frac{0.46}{0.098} = 4.69$$

and

$$C_c = \frac{(0.21)^2}{(0.46)(0.098)} = 0.98$$

General Comments

The diameter, D_{10}, is generally referred to as effective size. The effective size is used for several empirical correlations, such as coefficient of permeability. The coefficient of gradation, C_u, is a parameter which indicates the range of distribution of grain sizes in a given soil specimen. If C_u is relatively large, it indicates a well graded soil. If C_u is nearly equal to one, it means that the soil grains are of approximately equal size, and the soil may be referred to as a poorly graded soil.

Figure 4–5 shows the general nature of the grain-size distribution curves for a well graded and a poorly graded soil. In some instances, a soil may have a combination of two or more uniformly graded fractions, and this soil is referred to as gap graded. The grain-size distribution curve for a gap graded soil is also shown in Fig. 4–5.

The parameter C_c is also referred to as the *coefficient of curvature*. For sand, if C_u is greater than 6 and C_c is between 1 and 3, it is considered well graded. However, for a gravel to be well-graded, C_u should be greater than 4 and C_c must be between 1 and 3.

The D_{15} and D_{85} sizes are used for design of filters. The D_{50} size is used for correlation of the liquefaction potential of saturated granular soil during earthquakes.

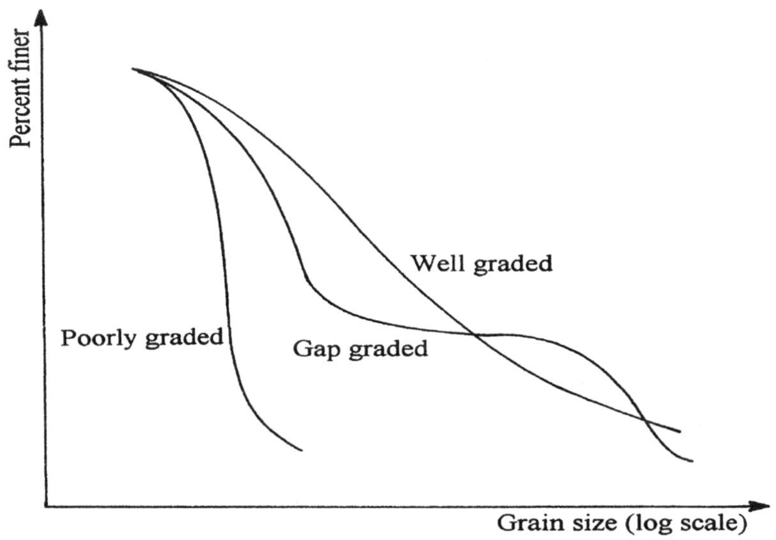

Figure 4–5. General nature of grain-size distribution of well graded, poorly graded and gap graded soil.

5

Hydrometer Analysis

Introduction

Hydrometer analysis is the procedure generally adopted for determination of the particle-size distribution in a soil for the fraction that is finer than No. 200 sieve size (0.075 mm). The lower limit of the particle-size determined by this procedure is about 0.001 mm.

In hydrometer analysis, a soil specimen is dispersed in water. In a dispersed state in the water, the soil particles will settle individually. It is assumed that the soil particles are spheres, and the velocity of the particles can be given by Stoke's law as

$$\upsilon = \frac{\gamma_s - \gamma_w}{18\eta} D^2 \tag{5.1}$$

where υ = velocity (cm/s)
γ_s = specific weight of soil solids (g/cm^3)
γ_w = unit weight of water (g/cm^3)
η = viscosity of water $\left(\frac{\text{g} \cdot \text{s}}{\text{cm}^2}\right)$
D = diameter of the soil particle

If a hydrometer is suspended in water in which soil is dispersed (Fig. 5–1), it will measure the specific gravity of the soil-water suspension at a depth L. The depth L is called the *effective depth*. So, at a time t minutes from the beginning of the test, the soil particles that settle beyond the zone of measurement (i.e., beyond the effective depth L) will have a diameter given by

$$\frac{L\ (\text{cm})}{t\ (\text{min}) \times 60} = \frac{(\gamma_s - \gamma_w)\ \text{g}/\text{cm}^3}{18\eta \left(\frac{\text{g} \cdot \text{s}}{\text{cm}^2}\right)} \left[\frac{D\ (\text{mm})}{10}\right]^2$$

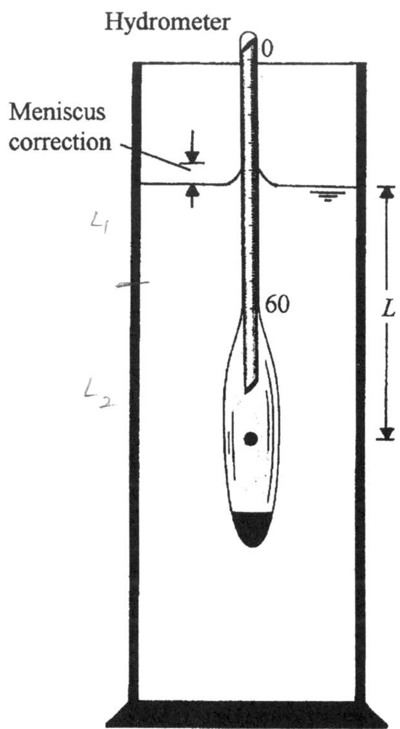

Figure 5–1. Hydrometer suspended in water in which the soil is dispersed.

$$D\,(\text{mm}) = \frac{10}{\sqrt{60}}\sqrt{\frac{18\eta}{(\gamma_s - \gamma_w)}}\sqrt{\frac{L}{t}} = A\sqrt{\frac{L\,(\text{cm})}{t\,(\text{min})}} \tag{5.2}$$

$$\text{where } A = \sqrt{\frac{1800\eta}{60(\gamma_s - \gamma_w)}} = \sqrt{\frac{30\eta}{(\gamma_s - \gamma_w)}} \tag{5.3}$$

In the test procedure described here, the *ASTM 152-H* type of hydrometer will be used. From Fig. 5–1 it can be seen that, based on the hydrometer reading (which increases from zero to 60 in the *ASTM 152-H* type of hydrometer), the value of *L* will change. The magnitude of *L* can be given as

$$L = L_1 + \frac{1}{2}\left(L_2 - \frac{V_B}{A_C}\right) \tag{5.4}$$

where L_1 = distance between the top of hydrometer bulb to the mark for a hydrometer reading. For a hydrometer reading of zero, L_1 = 10.5 cm. Also, for a hydrometer reading of 50 g/liter, L_1 = 2.3 cm. Thus, in general, for a given hydrometer reading

$$L_1 \text{ (cm)} = 10.5 - \left(\frac{10.5 - 2.3}{50}\right) \times \text{(hydrometer reading)}$$

L_2 = 14 cm
V_B = volume of the hydrometer bulb = 67.0 cm^3
A_c = cross-sectional area of the hydrometer cylinder = 27.8 cm^2

Based on Eq. (5.4), the variation of L with hydrometer reading is shown in Table 5–1.

For actual calculation purposes we also need to know the values of A given by Equation (5.3). An example of this calculation is shown below.

$$\gamma_s = G_s \gamma_w$$

where G_s = specific gravity of soil solids
Thus

$$A = \sqrt{\frac{30\eta}{(G_s - 1)\gamma_w}} \qquad (5.5)$$

For example, if the temperature of the water is 25°C,

$$\eta = 0.0911 \times 10^{-4} \left(\frac{\text{g} \cdot \text{s}}{\text{cm}^2}\right)$$

and $G_s = 2.7$

$$A = \sqrt{\frac{30(0.0911 \times 10^{-4})}{(2.7 - 1)(1)}} = 0.0127$$

The variations of A with G_s and the water temperature are shown in Table 5–2.

Table 5–1. Variation of L with hydrometer reading—*ASTM 152-H* hydrometer

Hydrometer reading	L (cm)	Hydrometer reading	L (cm)
0	16.3	26	12.0
1	16.1	27	11.9
2	16.0	28	11.7
3	15.8	29	11.5
4	15.6	30	11.4
5	15.5	31	11.2
6	15.3	32	11.1
7	15.2	33	10.9
8	15.0	34	10.7
9	14.8	35	10.6
10	14.7	36	10.4
11	14.5	37	10.2
12	14.3	38	10.1
13	14.2	39	9.9
14	14.0	40	9.7
15	13.8	41	9.6
16	13.7	42	9.4
17	13.5	43	9.2
18	13.3	44	9.1
19	13.2	45	8.9
20	13.0	46	8.8
21	12.9	47	8.6
22	12.7	48	8.4
23	12.5	49	8.3
24	12.4	50	8.1
25	12.2	51	7.9

The *ASTM 152-H* type of hydrometer is calibrated up to a reading of 60 at a tem- perature of 20°C for soil particles having a $G_s = 2.65$. A hydrometer reading of, say, 30 at a given time of a test means that there are 30 g of soil solids ($G_s = 2.65$) in suspension per 1000 cc of soil-water mixture at a temperature of 20°C at a depth where the specific gravity of the soil-water suspension is measured (i.e., L). From this measurement, we can determine the percentage of soil still in suspension at time t from the beginning of the test and all the soil particles will have diameters smaller than D calculated by Equation (5.2). However, in the actual experimental work, some corrections to the observed hydrometer readings need to be applied. They are as follows:

Table 5–2. Variation of A with G_s

G_s	Temperature (°C)						
	17	18	19	20	21	22	23
2.50	0.0149	0.0147	0.0145	0.0143	0.0141	0.0140	0.0138
2.55	0.0146	0.0144	0.0143	0.0141	0.0139	0.0137	0.0136
2.60	0.0144	0.0142	0.1040	0.0139	0.0137	0.0135	0.0134
2.65	0.0142	0.0140	0.0138	0.0137	0.0135	0.0133	0.0132
2.70	0.0140	0. 0138	0.1036	0.0134	0.0133	0.0131	0.0130
2.75	0.0138	0.0136	0.0134	0.0133	0.0131	0.0129	0.0128
2.80	0.0136	0.0134	0.0132	0.0131	0.0129	0.0128	0.0126

G_s	Temperature (°C)						
	24	25	26	27	28	29	30
2.50	0.0137	0.0135	0.0133	0.0132	0.0130	0.0129	0.0128
2.55	0.0134	0.0133	0.0131	0.0130	0.0128	0.0127	0.0126
2.60	0.0132	0.0131	0.0129	0.0128	0.0126	0.0125	0.0124
2.65	0.0130	0.0129	0.0127	0.0126	0.0124	0.0123	0.0122
2.70	0.0128	0.0127	0.0125	0.0124	0.0123	0.0121	0.0120
2.75	0.0126	0.0125	0.0124	0.0122	0.0121	0.0120	0.0118
2.80	0.0125	0.0123	0.0122	0.0120	0.0119	0.0118	0.0117

1. Temperature correction (F_T)—The actual temperature of the test may not be 20°C. The temperature correction (F_T) may be approximated as

$$F_T = -4.85 + 0.25T \text{ (for } T \text{ between 15°C and 28°C)} \tag{5.6}$$

where F_T = temperature correction to the observed reading (can be either positive or negative)
T = temperature of test in°C

2. Meniscus correction (F_m)—Generally, the upper level of the meniscus is taken as the reading during laboratory work (F_m is always positive).
3. Zero correction (F_z)— A deflocculating agent is added to the soil-distilled water suspension for performing experiments. This will change the zero reading (F_z can be either positive or negative).

Figure 5–2. Equipment for hydrometer test.

Equipment

1. *ASTM 152-H* hydrometer
2. Mixer
3. Two 1000-cc graduated cylinders
4. Thermometer
5. Constant temperature bath
6. Deflocculating agent
7. Spatula
8. Beaker
9. Balance
10. Plastic squeeze bottle
11. Distilled water
12. No. 12 rubber stopper

The equipment necessary (except the balance and the constant temperature bath) is shown in Fig. 5–2.

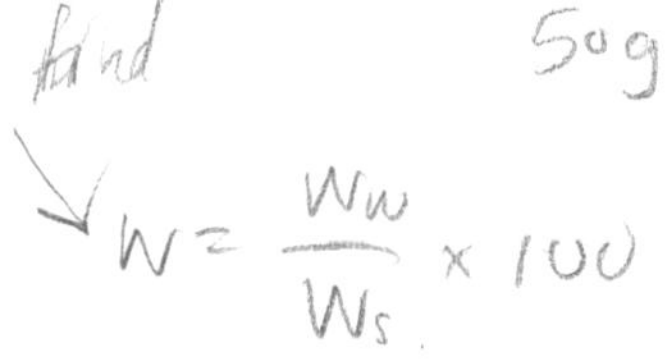

Procedure

Note: This procedure is used when more than 90 per cent of the soil is finer than No. 200 sieve.

1. Take 50 g of oven-dry, well-pulverized soil in a beaker.
2. Prepare a deflocculating agent. Usually a 4% solution of sodium hexametaphosphate (Calgon) is used. This can be prepared by adding 40 g of Calgon in 1000 cc of distilled water and mixing it thoroughly.
3. Take 125 cc of the mixture prepared in Step 2 and add it to the soil taken in Step 1. This should be allowed to soak for about 8 to 12 hours.
4. Take a 1000-cc graduated cylinder and add 875 cc of distilled water *plus* 125 cc of deflocculating agent in it. Mix the solution well.
5. Put the cylinder (from Step 4) in a constant temperature bath. Record the temperature of the bath, T (in °C).
6. Put the hydrometer in the cylinder (Step 5). Record the reading. (*Note:* The *top of the meniscus* should be read.) This is the zero correction (F_z), which can be $+ve$ or $-ve$. Also observe the meniscus correction (F_m).
7. Using a spatula, thoroughly mix the soil prepared in Step 3. Pour it into the mixer cup.

   Note: During this process, some soil may stick to the side of the beaker. Using the plastic squeeze bottle filled with distilled water, wash all the remaining soil in the beaker into the mixer cup.
8. Add distilled water to the cup to make it about two-thirds full. Mix it for about two minutes using the mixer.
9. Pour the mix into the second graduated 1000-cc cylinder. Make sure that all of the soil solids are washed out of the mixer cup. Fill the graduated cylinder with distilled water to bring the water level up to the 1000-cc mark.
10. Secure a No. 12 rubber stopper on the top of the cylinder (Step 9). Mix the soil-water well by turning the soil cylinder upside down several times.
11. Put the cylinder into the constant temperature bath next to the cylinder described in Step 5. Record the time immediately. This is cumulative time $t = 0$. Insert the hydrometer into the cylinder containing the soil-water suspension.
12. Take hydrometer readings at cumulative times $t = 0.25$ min., 0.5 min., 1 min., and 2 min. Always read the upper level of the meniscus.
13. Take the hydrometer out after two minutes and put it into the cylinder next to it (Step 5).
14. Hydrometer readings are to be taken at time $t = 4$ min., 8 min., 15 min., 30 min., 1 hr., 2 hr., 4 hr., 8 hr., 24 hr. and 48 hr. For each reading, insert the hydrometer into the cylinder containing the soil-water suspension about 30 seconds before the reading is due. After the reading is taken, remove the hydrometer and put it back into the cylinder next to it (Step 5).

Calculation

Refer to Table 5–4.

Column 2—These are observed hydrometer readings (R) corresponding to times given in Column 1.

Column 3—R_{cp} = corrected hydrometer reading for calculation of percent finer

$$= R + F_T - F_z \tag{5.7}$$

Column 4—Percent finer = $\frac{a\,R_{cp}}{W_s}(100)$

where W_s = dry weight of soil used for the hydrometer analysis

a = correction for specific gravity (since the hydrometer is calibrated for $G_s = 2.65$)

$$= \frac{G_s\,(1.65)}{(G_s - 1)2.65} \quad \text{(See Table 5-3)} \tag{5.8}$$

Table 5–3. Variation of *a* with G_s [Eq. 5.8]

G_s	a
2.50	1.04
2.55	1.02
2.60	1.01
2.65	1.00
2.70	0.99
2.75	0.98
2.80	0.97

Column 5—R_{cL} = corrected reading for determination of effective length = $R + F_m$ (5.9)

Column 6—Determine L (effective length) corresponding to the values of R_{cL} (Col. 5) given in Table 5–1.

Column 7—Determine A from Table 5–2.

Column 8— Determine $D\text{ (mm)} = A\sqrt{\frac{L\text{ (cm)}}{t\text{ (min)}}}$

Table 5–4. Hydrometer Analysis

Description of soil *Brown silty clay* Sample No.__________

Location ____

G_s *2.75* Hydrometer type *ASTM 152–H*

Dry weight of soil, W_s *50* g Temperature of test, T *28* °C

Meniscus correction, F_m *1* Zero correction, F_z *+7* Temperature correction, F_T *+2.15* [Eq. (5.6)]

Tested by____________________ Date____________________

Time (min) (1)	Hydrometer reading, R (2)	R_{cp} (3)	Percent finer, $\frac{a^* R_{cp}}{50} \times 100$ (4)	R_{cL} (5)	$L^{\dagger}$ (cm) (6)	$A^{\dagger}$ (7)	D (mm) (8)
0.25	51	46.15	90.3	52	7.8	0.0121	0.068
0.5	48	43.15	84.4	49	8.3		0.049
1	47	42.15	82.4	48	8.4		0.035
2	46	41.15	80.5	47	8.6		0.025
4	45	40.15	78.5	46	8.8		0.018
8	44	39.15	76.6	45	8.95		0.013
15	43	38.15	74.6	44	9.1		0.009
30	42	37.15	72.7	43	9.25		0.007
60	40	35.15	68.8	41	9.6		0.005
120	38	33.15	64.8	39	9.9		0.0035
240	34	29.15	57.0	35	10.5		0.0025
480	32	27.15	53.1	33	10.9		0.0018
1440	29	24.15	47.23	30	11.35		0.0011
2880	27	22.15	43.3	28	11.65		0.0008

* Table 5.3; † Table 5.1; ‡ Table 5.2

Graph

Plot a grain-size distribution graph on semi-log graph paper with percent finer (Col.4, Table 5–4) on the natural scale and D (Col. 8, Table 5–4) on the log scale. A sample calculation and the corresponding graph are shown in Table 5–4 and Fig. 5–3, respectively.

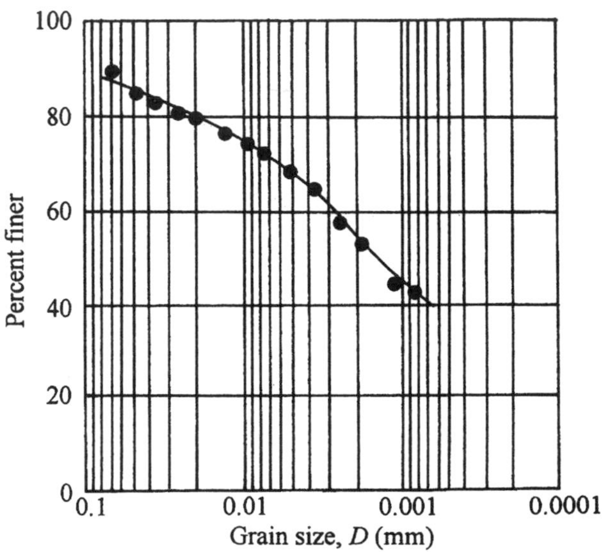

Figure 5–3. Plot of percent finer vs. grain size from the results given in Table 5–4.

Procedure Modification

When a smaller amount (less than about 90%) of soil is finer than No. 200 sieve size, the following modification to the above procedure needs to be applied.

1. Take an oven-dry sample of soil. Determine its weight (W_1).
2. Pulverize the soil using a mortar and rubber-tipped pestle, as described in Chapter 4.
3. Run a sieve analysis on the soil (Step 2), as described in Chapter 4.
4. Collect in the bottom pan the soil passing through No. 200 sieve.
5. Wash the soil retained on No. 200 sieve, as described in Chapter 4. Collect all the wash water and dry it in an oven.
6. Mix together the minus No. 200 portion from Step 4 and the dried minus No. 200 portion from Step 5.
7. Calculate the percent finer for the soil retained on No. 200 sieve and above (as shown in Table 4–1).
8. Take 50 g of the minus 200 soil (Step 6) and run a hydrometer analysis. (Follow Steps 1 through 14 as described previously.)

9. Report the calculations for the hydrometer analysis similar to that shown in Table 5–4. Note, however, that the percent finer now calculated (as in Col. 8 of Table 5–4) is *not the percent finer based on the total sample*. Calculate the percent finer based on the total sample as

$$P_T = (\text{Col. 8 of Table 5-4})\left(\frac{\text{percent passing No. 200 sieve}}{100}\right)$$

Percent passing No. 200 sieve can be obtained from Step 7 above.

10. Plot a combined graph for percent finer versus grain-size distribution obtained from *both the sieve analysis and the hydrometer analysis*. An example of this is shown in Fig. 5–4. From this plot, note that there is an overlapping zone. The percent finer calculated from the sieve analysis for a given grain size does not match that calculated from the hydrometer analysis. The grain sizes obtained from a sieve analysis are the least sizes of soil grains, and the grain sizes obtained from the hydrometer are the diameters of equivalent spheres of soil grains.

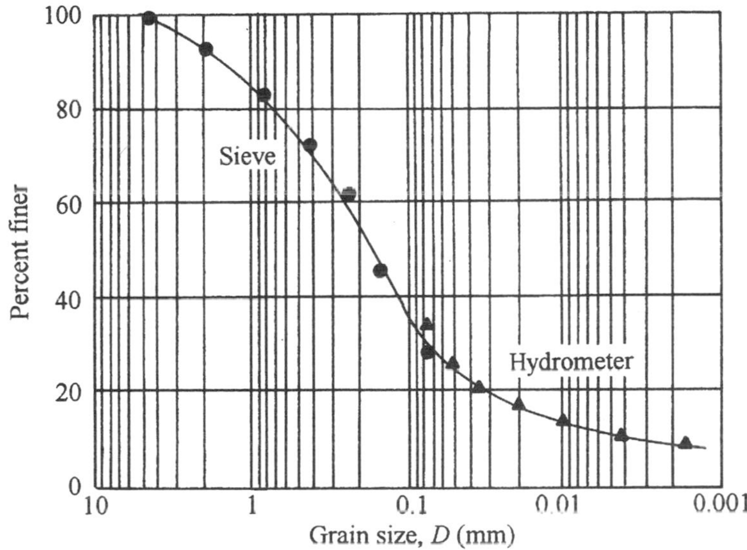

Figure 5–4. A grain-size distribution plot—combined results from sieve analysis and hydrometer analysis.

General Comments

A hydrometer analysis gives results from which the percent of soil finer than 0.002 mm in diameter can be estimated. It is generally accepted that the percent finer than 0.002 mm in size is clay or clay-size fractions. Most clay particles are smaller than 0.001 mm, and 0.002 mm is the upper limit. The presence of clay in a soil contributes to its plasticity.

6
Liquid Limit Test

Introduction

When a *cohesive soil* is mixed with an excessive amount of water, it will be in a somewhat *liquid state* and flow like a viscous liquid. However, when this viscous liquid is gradually dried, with the loss of moisture it will pass into a *plastic state*. With further reduction of moisture, the soil will pass into a semisolid and then into a solid state. This is shown in Fig. 6–1. The moisture content (in percent) at which the cohesive soil will pass from a liquid state to a plastic state is called the *liquid limit* of the soil. Similarly, the moisture contents (in percent) at which the soil changes from a plastic to a semisolid state and from a semisolid state to a solid state are referred to as the *plastic limit* and the *shrinkage limit*, respectively. These limits are referred to as the *Atterberg limits* (1911). In this chapter, the procedure to determine the liquid limit of a cohesive soil will be discussed.

Equipment

1. Casagrande liquid limit device
2. Grooving tool
3. Moisture cans
4. Porcelain evaporating dish

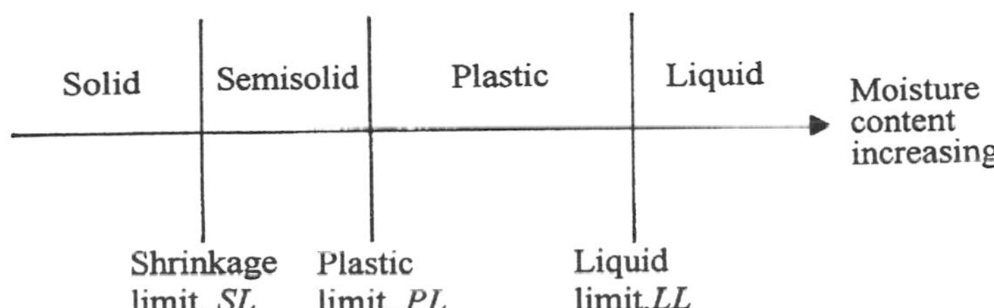

Figure 6–1. Atterberg limits.

5. Spatula
6. Oven
7. Balance sensitive up to 0.01 g
8. Plastic squeeze bottle
9. Paper towels

The equipment (except the balance and the oven) is shown in Fig. 6–2.

The Casagrande liquid limit device essentially consists of a brass cup that can be raised and dropped through a distance of 10 mm on a hard rubber base by a cam operated by a crank (see Fig. 6–3*a*). Fig. 6–3*b* shows a schematic diagram of a grooving tool.

Procedure

1. Determine the mass of three moisture cans (W_1).
2. Put about 250 g of air-dry soil, passed through No. 40 sieve, into an evaporating dish. Add water from the plastic squeeze bottle and mix the soil to the form of a uniform paste.
3. Place a portion of the paste in the brass cup of the liquid limit device. Using the spatula, smooth the surface of the soil in the cup such that the maximum depth of the soil is about 8 mm.
4. Using the grooving tool, cut a groove along the center line of the soil pat in the cup (Fig. 6–4*a*).
5. Turn the crank of the liquid limit device at the rate of about 2 revolutions per second. By this, the liquid limit cup will rise and drop through a vertical distance of 10 mm once for each revolution. The soil from two sides of the cup will begin to flow toward the center. Count the number of blows, *N*, for the groove in the soil to close through a distance of ½ in. (12.7 mm) as shown in Fig. 6–4*b*.

Figure 6–2. Equipment for liquid limit test.

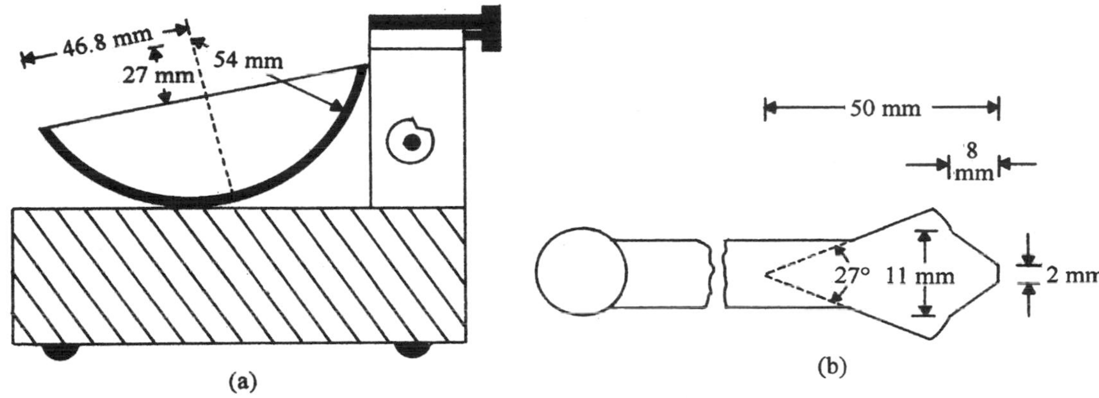

Figure 6–3. Schematic diagram of: (a) liquid limit device; (b) grooving tool.

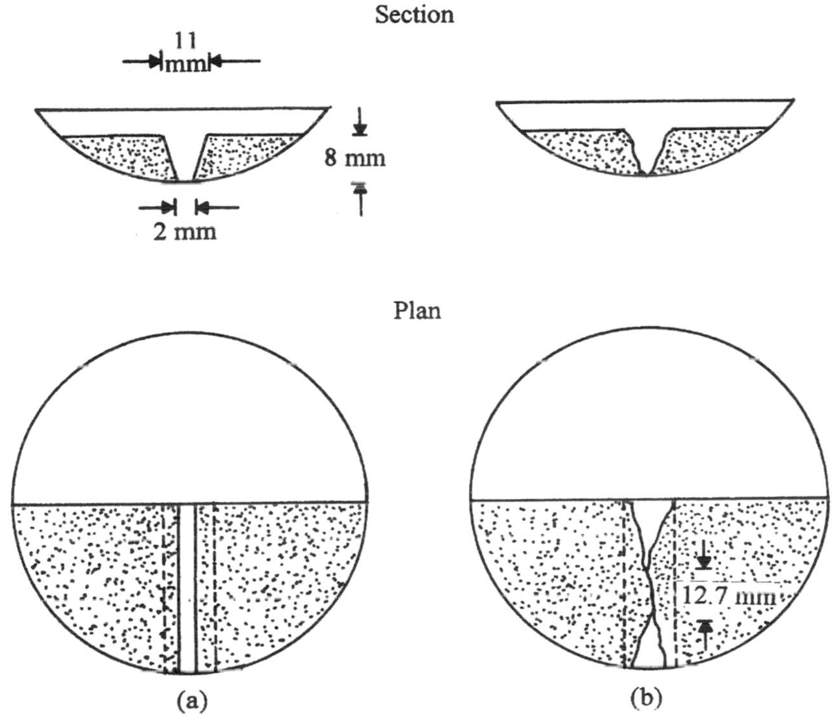

Figure 6–4. Schematic diagram of soil pat in the cup of the liquid limit device at (a) beginning of test, (b) end of test.

If N = about 25 to 35, collect a moisture sample from the soil in the cup in a moisture can. Close the cover of the can, and determine the mass of the can plus the moist soil (W_2).

Remove the rest of the soil paste from the cup to the evaporating dish. Use paper towels to thoroughly clean the cup.

If the soil is too dry, N will be more than about 35. In that case, remove the soil with the spatula to the evaporating dish. Clean the liquid limit cup thoroughly with paper towels. Mix the soil in the evaporating dish with more water, and try again.

If the soil is too wet, N will be less than about 25. In that case, remove the soil in the cup to the evaporating dish. Clean the liquid limit cup carefully with paper towels. Stir the soil paste with the spatula for some time to dry it up. The evaporating dish may be placed in the oven for a few minutes for drying also. *Do not* add dry soil to the wet-soil paste to reduce the moisture content for bringing it to the proper consistency. Now try again in the liquid limit device to get the groove closure of ½ in. (12.7 mm) between 25 and 35 blows.

6. Add more water to the soil paste in the evaporating dish and mix thoroughly. Repeat Steps 3, 4 and 5 to get a groove closure of ½ in. (12.7 mm) in the liquid limit device at a blow count N = 20 to 25. Take a moisture sample from the cup. Remove the rest of the soil paste to the evaporating dish. Clean the cup with paper towels.
7. Add more water to the soil paste in the evaporating dish and mix well. Repeat Steps 3, 4 and 5 to get a blow count N between 15 and 20 for a groove closure of ½ in. (12.7 mm) in the liquid limit device. Take a moisture sample from the cup.
8. Put the three moisture cans in the oven to dry to constant masses (W_3). (The caps of the moisture cans should be removed from the top and placed at the bottom of the respective cans in the oven.)

Calculation

Determine the moisture content for each of the three trials (Steps 5, 6 and 7) as

$$w\ (\%) = \frac{W_2 - W_3}{W_3 - W_1}(100) \tag{6.1}$$

Graph

Plot a semi-log graph between moisture content (arithmetic scale) versus number of blows, N (log scale). This will approximate a straight line, which is called the *flow curve*. From the straight line, determine the moisture content w (%) corresponding to 25 blows. This is the *liquid limit* of the soil.

The magnitude of the slope of the flow line is called the *flow index*, F_I, or

$$F_I = \frac{w_1\ (\%) - w_2\ (\%)}{\log N_2 - \log N_1} \tag{6.2}$$

Typical examples of liquid limit calculation and the corresponding graphs are shown in Table 6–1 and Fig. 6–5.

Table 6–1. Liquid Limit Test

Description of Soil *Gray silty clay* Sample No. *4*

Location

Tested by Date

Test No.	1	2	3
Can No.	*8*	*21*	*25*
Mass of can, W_1 (g)	*15.26*	*17.01*	*15.17*
Mass of can + moist soil, W_2 (g)	*29.30*	*31.58*	*31.45*
Mass of can + dry soil, W_3 (g)	*25.84*	*27.72*	*26.96*
Moisture content, w (%) $= \frac{W_2 - W_3}{W_3 - W_1} \times 100$	*32.7*	*36.04*	*38.1*
Number of blows, N	*35*	*23*	*17*

Liquid limit = *35.2*

Flow index = $\frac{(37 - 33.7)}{(\log 30 - \log 20)} = 18.74$

General Comments

Based on the liquid limit tests on several soils, the U.S. Army Waterways Experiment Station (1949) observed that the liquid limit, LL, of a soil can be approximately given by

$$LL = w_N\ (\%) \left(\frac{N}{25}\right)^{0.121} \tag{6.3}$$

where w_N (%) = moisture content, in percent, for ½ in. (12.7 mm) groove closure in the liquid limit device at N number of blows

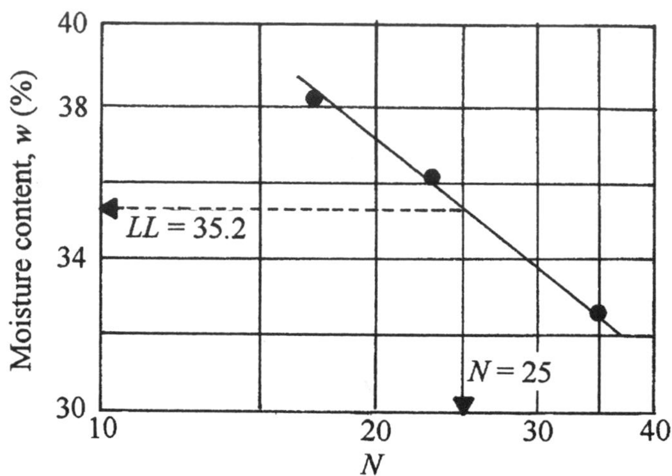

Figure 6–5. Plot of moisture content (%) vs. number of blows for the liquid limit test results reported in Table 6–1.

ASTM also recommends this equation for determining the liquid limit of soils (ASTM designation D-4318). However, the value of w_N should correspond to an N value between 20 and 30. Following are the values of $(N/25)^{0.121}$ for various values of N.

N	$\left(\frac{N}{25}\right)^{0.121}$	N	$\left(\frac{N}{25}\right)^{0.121}$
20	0.973	26	1.005
21	0.979	27	1.009
22	0.985	28	1.014
23	0.990	29	1.018
24	0.995	30	1.022
25	1.000		

The presence of clay contributes to the plasticity of soil. The liquid limit of a soil will change depending on the amount and type of clay minerals present in it. Following are the approximate ranges for the liquid limit of some clay minerals

Clay mineral	*LL*
Kaolinite	35–100
Illite	55–120
Montmorillonite	100–800

7
Plastic Limit Test

Introduction

The fundamental concept of *plastic limit* was introduced in the introductory section of the preceding chapter (see Fig. 6–1). Plastic limit is defined as the moisture content, in percent, at which a cohesive soil will change from a *plastic state* to a *semisolid state*. In the laboratory, the *plastic limit* is defined as the moisture content (%) at which a thread of soil will just crumble when rolled to a diameter of ⅛-in. (3.18 mm). This test might be seen as somewhat arbitrary and, to some extent, the result may depend on the person performing the test. With practice, however, fairly consistent results may be obtained.

Equipment

1. Porcelain evaporating dish
2. Spatula
3. Plastic squeeze bottle with water
4. Moisture can
5. Ground glass plate
6. Balance sensitive up to 0.01 g

Procedure

1. Put approximately 20 grams of a representative, air-dry soil sample, passed through No. 40 sieve, into a porcelain evaporating dish.
2. Add water from the plastic squeeze bottle to the soil and mix thoroughly.
3. Determine the mass of a moisture can in grams and record it on the data sheet (W_1).
4. From the moist soil prepared in Step 2, prepare several ellipsoidal-shaped soil masses by squeezing the soil with your fingers.
5. Take one of the ellipsoidal-shaped soil masses (Step 4) and roll it on a ground glass

plate using the palm of your hand (Fig. 7–1). The rolling should be done at the rate of about 80 strokes per minute. Note that one complete backward and one complete forward motion of the palm constitute a stroke.

Figure 7–1. An ellipsoidal soil mass is being rolled into a thread on a glass plate.

6. When the thread is being rolled in Step 5 reaches ⅛-in. (3.18 mm) in diameter, break it up into several small pieces and squeeze it with your fingers to form an ellipsoidal mass again.
7. Repeat Steps 5 and 6 until the thread crumbles into several pieces when it reaches a diameter of ⅛-in. (3.18 mm).It is possible that a thread may crumble at a diameter larger than ⅛-in. (3.18 mm) during a given rolling process, whereas it did not crumble at the same diameter during the immediately previous rolling.
8. Collect the small crumbled pieces in the moisture can put the cover on the can.
9. Take the other ellipsoidal soil masses formed in Step 4 and repeat Steps 5 through 8.
10. Determine the mass of the moisture can plus the wet soil (W_2) in grams. Remove the cap from the top of the can and place the can in the oven (with the cap at the bottom of the can).
11. After about 24 hours, remove the can from the oven and determine the mass of the can plus the dry soil (W_3) in grams.

Calculations

$$\text{Plastic limit} = \frac{\text{mass of moisture}}{\text{mass of dry soil}} = \frac{W_2 - W_3}{W_3 - W_1}(100) \tag{7.1}$$

The results may be presented in a tabular form as shown in Table 7–1. If the liquid limit of the soil is known, calculate the *plasticity index, PI*, as

$$PI = LL - PL \tag{7.2}$$

Table 7–1. Plastic Limit Test

Description of soil *Gray clayey silt* Sample No. *3*

Location ______

Tested by ______ Date ______

Can No.	*103*
Mass of can, W_1 (g)	*13.33*
Mass of can + moist soil, W_2 (g)	*23.86*
Mass of can + dry soil, W_3 (g)	*22.27*
$PL = \frac{W_2 - W_3}{W_3 - W_1} \times 100$	*17.78*

Plasticity index, $PI = LL - PL =$ *34 - 17.78* = *16.28*

General Comments

The liquid limit and the plasticity index of cohesive soils are important parameters for classifiction purposes. The engineering soil classification systems are described in Chapter 9. The plasticity index is also used to determine the activity, A, of a clayey soil which is defined as

$$A = \frac{PI}{(\text{\% of clay - size fraction, by weight})}$$

Following are typical values of PI of several clay minerals.

Clay minerals	*PI*
Kaolinite	20–40
Illite	35–50
Montmorillonite	50–100

8
Shrinkage Limit Test

Introduction

The fundamental concept of *shrinkage limit* was presented in Fig. 6–1. A saturated clayey soil, when gradually dried, will lose moisture and, subsequently, there will be a reduction in the volume of the soil mass. During the drying process, a condition will be reached when any further drying will result in a reduction of moisture content without any decrease in volume (Fig.8–1). The moisture content of the soil, in percent, at which the decrease in soil volume ceases is defined as the *shrinkage limit*.

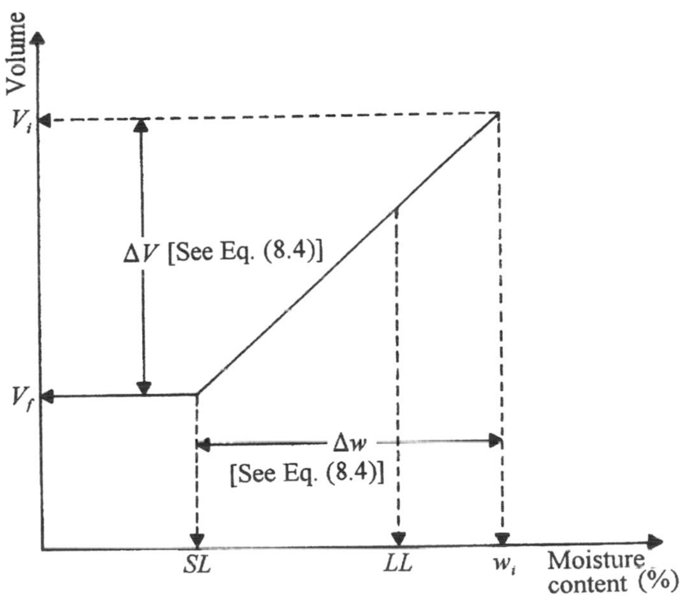

Figure 8–1. Definition of shrinkage limit.

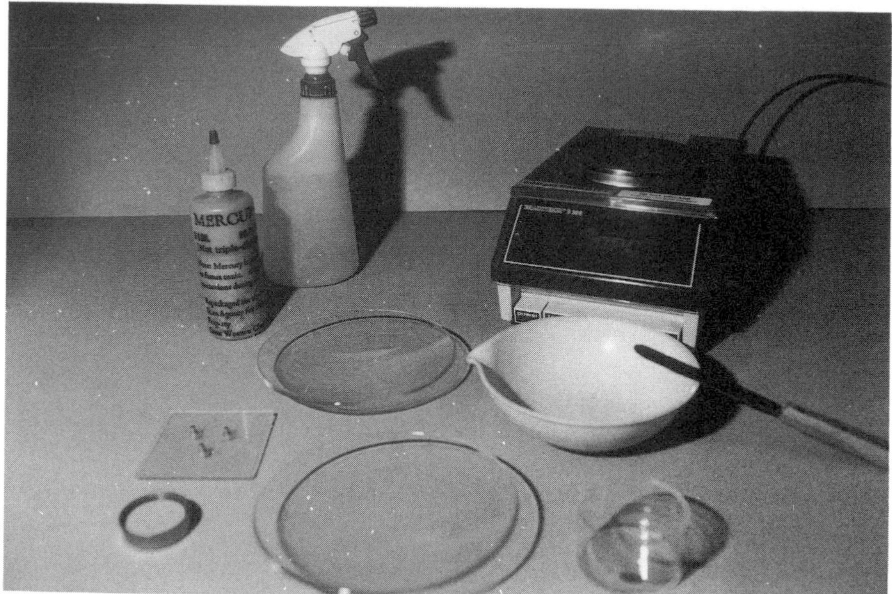

Figure 8–2. Equipment needed for determination of shrinkage limit.

Equipment

1. Shrinkage limit dish [usually made of porcelain, about 1.75 in. (44.4 m) in diameter and 0.5 in. (12.7 mm) high]
2. A glass cup [2.25 in. (57.15 mm) in diameter and 1.25 in. (31.75 mm) high]
3. Glass plate with three prongs
4. Porcelain evaporating dish about 5.5 in. (139.7 mm) diameter
5. Spatula
6. Plastic squeeze bottle with water
7. Steel straight edge
8. Mercury
9. Watch glass
10. Balance sensitive to 0.01 g

The above required equipment is shown in Fig. 8–2.

Procedure

1. Put about 80 to 100 grams of a representative air dry soil, passed through No. 40 sieve, into an evaporating dish.
2. Add water to the soil from the plastic squeeze bottle and mix it thoroughly into the form of a creamy paste. Note that the moisture content of the paste should be above the liquid limit of the soil to ensure full saturation.
3. Coat the shrinkage limit dish lightly with petroleum jelly and then determine the mass of the coated dish (W_1) in grams.
4. Fill the dish about one-third full with the soil paste. Tap the dish on a firm surface

so that the soil flows to the edges of the dish and no air bubbles exist.

5. Repeat Step 4 until the dish is full.
6. Level the surface of the soil with the steel straight edge. Clean the sides and bottom of the dish with paper towels.
7. Determine the mass of the dish plus the wet soil (W_2) in grams.
8. Allow the dish to air dry (about 6 hours) until the color of the soil pat becomes lighter. Then put the dish with the soil into the oven to dry.
9. Determine the mass of the dish and the oven-dry soil pat (W_3) in grams.
10. Remove the soil pat from the dish.
11. In order to find the volume of the shrinkage limit dish (V_i), fill the dish with mercury. (*Note:* The dish should be placed on a watch glass.) Use the three-pronged glass plate and level the surface of the mercury in the dish. The excess mercury will flow into the watch glass. Determine the mass of mercury in the dish (W_4) in grams..
12. In order to determine the volume of the dry soil pat (V_f), fill the glass cup with mercury. (The cup should be placed on a watch glass.) Using the three-pronged glass plate, level the surface of the mercury in the glass cup. Remove the excess mercury on the watch glass. Place the dry soil pat on the mercury in the glass cup. The soil pat will float. Now, using the three-pronged glass plate, slowly push the soil pat into the mercury until the soil pat is completely submerged (Fig. 8–3). The displaced mercury will flow out of the glass cup and will be collected on the watch glass. Determine the mass of the displaced mercury on the watch glass (W_5) in grams.

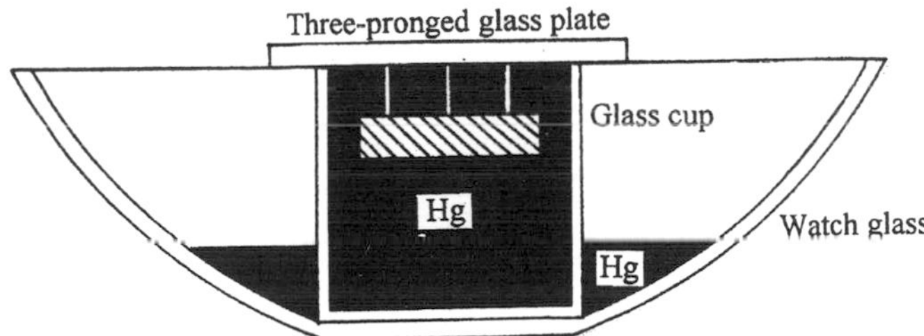

Figure 8–3. Determination of the volume of the soil pat (Step 12).

Calculation

1. Calculate the initial moisture content of the soil at molding.

$$w_i\ (\%) = \frac{(W_2 - W_3)}{(W_3 - W_1)} \times 100 \tag{8.1}$$

2. Calculate the change in moisture content (%) before the volume reduction ceased (refer to Fig. 8–1).

$$\Delta w_i\ (\%) = \frac{(V_i - V_f)\rho_w}{\text{mass of dry soil pat}} = \frac{(W_4 - W_5)}{13.6\ (W_3 - W_1)}(100) \tag{8.2}$$

where ρ_w = density of water = 1 g/cm^3

3. Calculate the shrinkage limit.

$$SL = w_i - \frac{(W_4 - W_5)}{13.6(W_3 - W_1)}(100) \tag{8.3}$$

Note that W_4 and W_5 are in grams, and the specific gravity of the mercury is 13.6. A sample calculation is shown in Table 8–1.

Table 8–1. Shrinkage Limit Test

Description of soil *Dark brown clayey silt* Sample No. *8*

Location *Westwind Boulevard*

Tested by ________ Date ________

Test No.	*1*	
Mass of coated shrinkage limit dish, W_1 (g)	*12.34*	
Mass of dish + wet soil, W_2 (g)	*40.43*	
Mass of dish + dry soil, W_3 (g)	*33.68*	
$w_i\ (\%) = \frac{(W_2 - W_3)}{(W_3 - W_1)} \times 100$	*31.63*	
Mass of mercury to fill the dish, W_4 (g)	*198.83*	
Mass of mercury displaced by soil pat, W_5 (g)	*150.30*	
$\Delta w_i\ (\%) = \frac{(W_4 - W_5)}{(13.6)(W_3 - W_1)} \times 100$	*16.72*	
$SL = w_i - \frac{(W_4 - W_5)}{13.6(W_3 - W_1)}(100)$	*14.91*	

General Comments

The ratio of the liquid limit to the shrinkage limit (*LL*/*SL*) of a soil gives a good idea about the shrinkage properties of the soil. If the ratio of *LL*/*SL* is large, the soil in the field may undergo undesirable volume change due to change in moisture. New foundations constructed on these soils may show cracks due to shrinking and swelling of the soil that result from seasonal moisture change.

Another parameter called *shrinkage ratio* (*SR*) may also be determined from the shrinkage limit test. Referring to Fig. 8–1

$$SR = \frac{\Delta V / V_f}{\Delta w / W_s} = \frac{\Delta V / V_f}{(\Delta V \rho_w) / W_s} = \frac{W_s}{\rho_w V_f} \tag{8.4}$$

where W_s = weight of the dry soil pat
$= W_3 - W_1$

If W_s is in grams, V_f is in cm³ and $\rho_w = 1$ g/cm³. So

$$SR = \frac{W_s}{V_f} \tag{8.5}$$

The shrinkage ratio gives an indication of the volume change with change in moisture content.

9

Engineering Classification of Soils

Introduction

Soils are widely varied in their grain-size distribution (Chapters 4 and 5). Also, depending on the type and quantity of clay minerals present, the plastic properties of soils (Chapters 6, 7 and 8) may be very different. Various types of engineering works require the identification and classification of soil in the field. In the design of foundations and earth-retaining structures, construction of highways, and so on, it is necessary for soils to be arranged in specific groups and/or subgroups based on their grain-size distribution and plasticity. The process of placing soils into various groups and/or subgroups is called *soil classification.*

For engineering purposes, there are two major systems that are presently used in the United States. They are: (i) the *American Association of State Highway and Transportation Officials (AASHTO) Classification System* and (ii) the *Unified Classification System*. These two systems will be discussed in this chapter.

American Association of State Highway and Transportation Officials (AASHTO) System of Classification

The AASHTO classification system was originally initiated by the Highway Research Board (now called the Transportation Research Board) in 1943. This classification system has under-gone several changes since then. This system is presently used by federal, state, and county highway departments in the United States. In this soil classification system, soils are generally placed in seven major groups: *A-1, A-2, A-3, A-4, A-5, A-6* and *A-7*. Group *A-1* is divided into two subgroups: *A-1-a* and *A-1-b*. Group *A-2* is divided into four subgroups: *A-2-4, A-2-5, A-2-6* and *A-2-7*. Soils under group *A-7* are also divided into two subgroups: *A-7-5* and *A-7-6*. This system is also presently included in ASTM under test designation D-3284.

Along with the soil groups and subgroups discussed above, another factor called the *group index* (GI) is also included in this system. The importance of group index can be explained as follows. Let us assume that two soils fall under the same group; however, they may have different values of GI. The soil that has a lower value of group index is likely to perform better as a highway subgrade material.

The procedure for classifying soil under the AASHTO system is outlined below.

Step-by-Step Procedure for AASHTO Classification

1. Determine the percentage of soil passing through U.S. No. 200 sieve (0.075 mm opening).
   If 35% or less passes No. 200 sieve, it is a coarse-grained material. Proceed to Steps 2 and 4.
   If more than 35% passed No. 200 sieve, it is a fine-grained material (i.e., silty or clayey material). For this, go to Steps 3 and 5.

Determination of Groups or Subgroups

2. For coarse-grained soils, determine the percent passing U.S. sieve Nos. 10, 40 and 200 and, additionally, the liquid limit and plasticity index. Then proceed to Table 9.1. Start from the top line and compare the known soil properties with those given in the table (Columns 2 through 6). Go down one line at a time until a line is found for which all the properties of the desired soil matches. The soil group (or subgroup) is determined from Column 1.
3. For fine-grained soils, determine the liquid limit and the plasticity index. Then go to Table 9.2. Start from the top line. By matching the soil properties from Columns 2, 3 and 4, determine the proper soil group (or subgroup).

Determination of Group Index

4. To determine the group index (GI) of coarse-grained soils, the following rules need to be observed.
   a. GI for soils in groups (or subgroups) *A-1-a, A-1-b, A-2-4, A-2-5* and *A-3* is zero.
   b. For GI in soils of groups *A-2-6* and *A-2-7*, use the following equation:

$$GI = 0.01(F_{200} - 15)(PI - 10) \tag{9.1}$$

   where F_{200} = percent passing No. 200 sieve
   PI = plasticity index
   If the GI comes out negative, round it off to zero. If the GI is positive, round it off to the nearest whole number.

Table 9–1. AASHTO Classification for Coarse-Grained Soils

Soil group (1)		Grain size: Passing No. 10 sieve (2)	Grain size: Passing No. 40 sieve (3)	Grain size: Passing No. 200 sieve (4)	Liquid limit* (5)	Plasticity index* (6)	Material type (7)	Subgrade rating (8)
A-1	*A-1-a*	50 max.	30 max.	15 max.		6 max	Stone fragments, gravel and sand	Excellent to good
	A-1-b		50 max.	25 max.		6 max.		
A-3			51 min.	10 max.		Nonplastic	Fine sand	
A-2	*A-2-4*			35 max.	40 max.	10 max.	Silty and clayey gravel and sand	
	A-2-5			35 max.	41 min.	10 max.		
	A-2-6			35 max.	40 max.	11 min.		
	A-2-7			35 max.	41 min.	11 min.		

* Based on the fraction passing No. 40 sieve

Table 9–2. AASHTO Classification for Fine-Grained Soils

Soil group (1)		Passing No. 200 sieve (2)	Liquid limit* (3)	Plasticity index* (4)	Material type (5)	Subgrade rating (6)
A-4		36 min.	40 max.	10 max.	Silty soil	Fair to poor
A-5		36 min.	41 min.	10 max.	Silty soil	Fair to poor
A-6		36 min.	40 max.	11 min.	Clayey soil	Fair to poor
A-7	*A-7-5*	36 min.	41 min.	11 min. and $PI \leq LL - 30$	Clayey soil	Fair to poor
	A-7-6	36 min.	41 min.	11 min. and $PI > LL - 30$	Clayey soil	Fair to poor

* Based on the fraction passing U.S. No. 40 sieve

5. For obtaining the GI of coarse-grained soils, use the following equation:

$$GI = (F_{200} - 35)[0.2 + 0.005(LL - 40)] + 0.01(F_{200} - 15)(PI - 10) \qquad (9.2)$$

If the GI comes out negative, round it off to zero. However, if it is positive, round it off to the nearest whole number.

Expression for Soil Classification

6. The final classification of a soil is given by first writing down the group (or subgroup) followed by the group index in parenthesis.

General

Figure 9–1 shows the range of *PI* and *LL* for soil groups *A-2-4, A-2-5, A-2-6, A-2-7, A-4, A-5, A-6, A-7-5* and *A-7-6*.

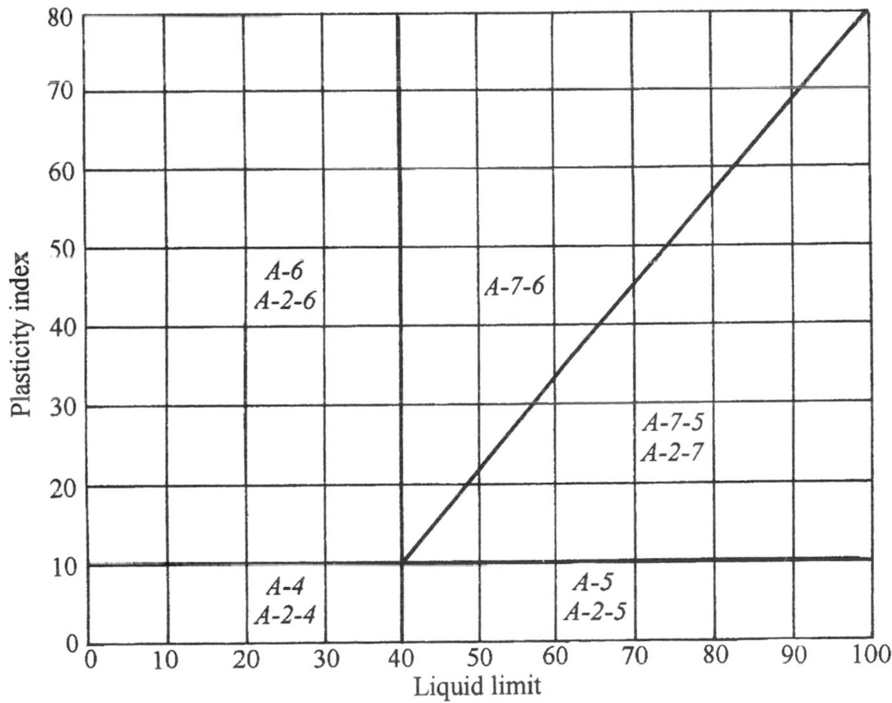

Figure 9–1. Liquid limit and plasticity index for nine AASHTO soil groups.

Example 9–I

The following are the characteristics of two soils. Classify the soils according to the AASHTO system.

Soil A:

Percent passing No. 4 sieve	=	98
Percent passing No. 10 sieve	=	90
Percent passing No. 40 sieve	=	76
Percent passing No. 200 sieve	=	34
Liquid limit	=	38
Plastic limit	=	26

Soil B:

Percent passing No. 4 sieve	=	100
Percent passing No. 10 sieve	=	98
Percent passing No. 40 sieve	=	86
Percent passing No. 200 sieve	=	58
Liquid limit	=	49
Plastic limit	=	28

Solution

Soil A:

1. The soil has 34% (which is less than 35%) passing through No. 200 sieve. So this is a coarse-grained soil.
2. For this soil, the liquid limit = 38.

   From Equation (7.2), plasticity index, $PI = LL - PL = 38 - 24 = 12$.

   From Table 9–1, by matching, the soil is found to belong to subgroup *A-2-6*.
3. From Equation (9.1)

   $$GI = 0.01(F_{200} - 15)(PI - 10)$$
   $$= 0.01(34 - 15)(12 - 10) = (0.01)(19)(2)$$
   $$= 0.38 \approx 0$$
4. So, the soil can be classified as *A-2-6(0)*.

Soil B:

1. The soil has 58% (which is more than 35%) passing through No.200 sieve. So this is a fine-grained soil.
2. The liquid limit of the soil is 49.

   From Equation (7.2), plasticity index, $PI = LL - PL = 49 - 28 = 21$.
3. From Table 9–2, the soil is either *A-7-5* or *A-7-6*. However, for this soil

   $$PI = 21 > LL - 30 = 49 - 30 = 19.$$

   So this soil is *A-7-6*.

4. From Equation (9–2)

$$GI = (F_{200} - 35)[0.2 + 0.005(LL - 40)] + 0.01(F_{200} - 15)(PI - 10)$$
$$= (58 - 35)[0.2 + 0.005(49 - 40)] + 0.01(58 - 15)(21 - 10)$$
$$= 5.64 + 4.73 = 10.37 \approx 10$$

5. So the soil is classified as *A-7-6(10)*.

Unified Classification System

This classification system was originally developed in 1942 by Arthur Casagrande for airfield construction during World War II. This work was conducted on behalf of the U.S. Army Corps of Engineers. At a later date, with the cooperation of the United States Bureau of Reclamation, the classification was modified. More recently, the American Society of Testing and Materials (ASTM) introduced a more definite system for group name of soils. In the pre-sent form, it is widely used by foundation engineers all over the world. Unlike the AASHTO system, the Unified system uses symbols to represent the soil types and the index properties of the soil. They are as follows:

Symbol	Soil type	Symbol	Index property
G	Gravel	*W*	Well-graded (for grain-size distribution)
S	Sand		
		P	Poorly-graded (for grain-size distribution)
M	Silt		
C	Clay	*L*	Low to medium plasticity
O	Organic silts and clays	*H*	High plasticity
Pt	Highly organic soil and peat		

Soil groups are developed by combining symbols for two categories listed above, such as *GW*, *SM*, and so forth.

Step-by-Step Procedure for Unified Classification System

1. If it is peat (i.e., primarily organic matter, dark in color, and has organic odor), classify it as *Pt* by visual observation. For all other soils, determine the percent of soil passing through U.S. No. 200 sieve (F_{200}).
2. Determine the percent retained on U.S. No. 200 sieve (R_{200}) as

$$R_{200} = 100 - \underset{\substack{\uparrow \\ \text{(nearest whole number)}}}{F_{200}} \tag{9.3}$$

3. If R_{200} is greater than 50%, it is a coarse-grained soil. However, if R_{200} is less than or equal to 50%, it is a fine-grained soil. For the case where $R_{200} \leq 50\%$ (i.e., fine-grained soil), go to Step 4. If $R_{200} > 50\%$, go to Step 5.
4. For fine-grained soils (i.e., $R_{200} \leq 50\%$, determine if the soil is organic or inorganic in nature.
   a. If the soil is organic, the group symbol can be *OH* or *OL*. If the soil is inorganic, the group symbol can be *CL, ML, CH, MH,* or *CL-ML*.
   b. Determine the percent retained on U.S. No. 4 sieve (R_4) as

$$R_4 = 100 - \underset{\substack{\uparrow \\ \text{(nearest whole number)}}}{F_4} \tag{9.4}$$

   where F_4 = percent finer than No. 4 sieve
   Note that R_4 is the percent of gravel fraction in the soil (*GF*), so

$$GF = R_4 \tag{9.5}$$

   c. Determine the percent of sand fraction in the soil (*SF*), or

$$SF = R_{200} - GF \tag{9.6}$$

   d. For inorganic soils, determine the liquid limit (*LL*) and the plasticity index (*PI*). Go to Step 4e. For organic soils, determine the liquid limit (not oven dried), LL_{NOD}; the liquid limit (oven dried), LL_{OD}; and the plasticity index (not oven dried), PI_{NOD}. Go to Step 4f.
   e. With known values of R_{200}, *GF, SF, SF/GF, LL* and *PI*, use Table 9–3 to obtain group symbols and group names of inorganic soils.
   f. With known values of LL_{NOD}, LL_{OD}, PI_{NOD}, R_{200}, *GF, SF* and *SF/GF*, use Table 9–4 to obtain group symbols and group names of organic soils.

   Figure 9–2 shows a plasticity chart with group symbols for fine-grained soils.
5. For coarse-grained soils:
   a. If $R_4 > 0.5R_{200}$, it is a gravelly soil. These soils may have the following group symbols:

      GW *GW-GM*
      GP *GW-GC*
      GM *GP-GM*
      GC *GP-GC*
      GC-GM

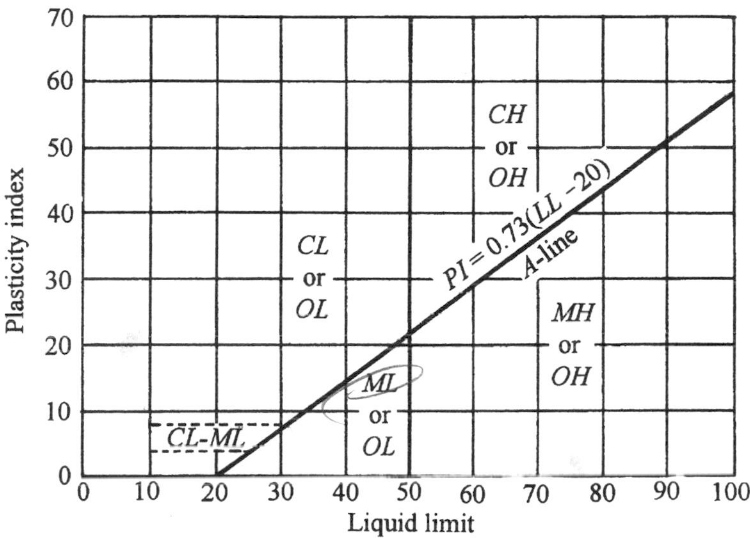

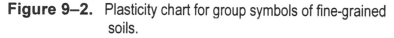

Figure 9–2. Plasticity chart for group symbols of fine-grained soils.

Determine the following:

(1) F_{200}

(2) Uniformity coefficient, $C_u = D_{60}/D_{10}$ (see Chapter 4)

(3) Coefficient of gradation, $C_c = D_{30}^2/(D_{60} \times D_{10})$

(4) LL (of minus No. 40 sieve)

(5) PI (of minus No. 40 sieve)

(6) SF [based on Equations (9.3), (9.4), (9.5) and (9.6)]

Go to Table 9–5 to obtain group symbols and group names.

b. If $R_4 \le 0.5R_{200}$, it is a sandy soil. These soils may have the following group symbols:

SW	*SW-SM*
SP	*SW-SC*
SM	*SP-SM*
SC	*SP-SC*
	SM-SC

Table 9–3. Unified Classification of Fine-Grained Inorganic Soils
(*Note:* The group names are based on ASTM D-2487.)

Criteria for group symbol	Group symbol	Criteria for group name				Group name
		R_{200}	*SF/GF*	*GF*	*SF*	
$LL < 50$, $PI > 7$, and $PI \geq 0.73(LL - 20)$	*CL*	<15	—	—	—	Lean clay
		15 to 29	≥1	—	—	Lean clay with sand
			<1	—	—	Lean clay with gravel
		≥30	≥1	<15	—	Sandy lean clay
			≥1	≥15	—	Sandy lean clay with gravel
			<1		<15	Gravelly lean clay
			<1		≥15	Gravelly lean clay with sand
$LL < 50$, $PI < 4$, or $PI < 0.73(LL - 20)$	*ML*	<15	—	—	—	Silt
		15 to 29	≥1	—	—	Silt with sand
			<1	—	—	Silt with gravel
		≥30	≥1	<15	—	Sandy silt
			≥1	≥15	—	Sandy silt with gravel
			<1		<15	Gravelly silt
			<1		≥15	Gravelly silt with sand
$LL < 50$, $4 \leq PI \leq 7$, and $PI \geq 0.73(LL - 20)$	*CL-ML*	<15	—	—	—	Silty clay
		15 to 29	≥1	—	—	Silty clay with sand
			<1	—	—	Silty clay with gravel
		≥30	≥1	<15	—	Sandy silty clay
			≥1	≥15	—	Sandy silty clay with gravel
			<1		<15	Gravelly silty clay
			<1		≥15	Gravelly silty clay with sand

continued

Table 9–3. *continued*

Criteria for group symbol	Group symbol	Criteria for group name				Group name
		R_{200}	SF/GF	GF	SF	
$LL \geq 50$, and $PI \geq 0.73(LL - 20)$	CH	<15	—	—	—	Fat clay
		15 to 29	≥1	—	—	Fat clay with sand
			<1	—	—	Fat clay with gravel
		≥30	≥1	<15	—	Sandy fat clay
			≥1	≥15	—	Sandy fat clay with gravel
			<1		<15	Gravelly fat clay
			<1		≥15	Gravelly fat clay with sand
$LL \geq 50$ and $PI < 0.73(LL - 20)$	MH	<15	—	—	—	Elastic silt
		15 to 29	≥1	—	—	Elastic silt with sand
			<1	—	—	Elastic silt with gravel
		≥30	≥1	<15	—	Sandy elastic silt
			≥1	≥15	—	Sandy elastic silt with gravel
			<1		<15	Gravelly elastic silt
			<1		≥15	Gravelly elastic silt with sand

Table 9–4. Unified Classification of Fine-Grained Organic Soils (*Note:* The group names are based on ASTM D-2487.)

Criteria for group symbol	Group symbol	Criteria for group name					Group name
		Plasticity index	R_{200}	*SF/GF*	*GF*	*SF*	
$LL_{NOD} <$ 50 and $\frac{LL_{OD}}{LL_{NOD}} < 0.75$	*OL*	$PI_{NOD} \geq 4$ and $PI_{NOD} \geq 0.73 \times (LL_{NOD} - 20)$	<15	—	—	—	Organic clay
			15 to 29	≥1	—	—	Organic clay with sand
				<1	—	—	Organic clay with gravel
			≥30	≥1	<15	—	Sandy organic clay
				≥1	≥15	—	Sandy organic clay with gravel
				<1		<15	Gravelly organic clay
				<1		≥15	Gravelly organic clay with sand
		$PI_{NOD} < 4$ and $PI_{NOD} < 0.73 \times (LL_{NOD} - 20)$	<15	—	—	—	Organic silt
			15 to 29	≥1	—	—	Organic silt with sand
				<1	—	—	Organic silt with gravel
			≥30	≥1	<15	—	Sandy organic silt
				≥1	≥15	—	Sandy organic silt with gravel
				<1		<15	Gravelly organic silt
				<1		≥15	Gravelly organic silt with sand

Table 9–4. *continued*

Criteria for group symbol	Group symbol	Criteria for group name					Group name
		Plasticity index	R_{200}	*SF/GF*	*GF*	*SF*	
$LL_{NOD} \geq$ 50, and $\frac{LL_{OD}}{LL_{NOD}} < 0.75$	*OH*	$PI_{NOD} \geq 0.73 \times (LL_{NOD} - 20)$	<15	—	—	—	Organic clay
			15 to 29	≥1	—	—	Organic clay with sand
				<1	—	—	Organic clay with gravel
			≥30	≥1	<15	—	Sandy organic clay
				≥1	≥15	—	Sandy organic clay with gravel
				<1		<15	Gravelly organic clay
				<1		≥15	Gravelly organic clay with sand
		$PI_{NOD} < 0.73 \times (LL_{NOD} - 20)$	<15	—	—	—	Organic silt
			15 to 29	≥1	—	—	Organic silt with sand
				<1	—	—	Organic silt with gravel
			≥30	≥1	<15	—	Sandy organic silt
				≥1	≥15	—	Sandy organic silt with gravel
				<1		<15	Gravelly organic silt
				<1		≥15	Gravelly organic silt with sand

Table 9–5. Unified Classification of Gravelly Soils ($R_4 > 0.5R_{200}$) (*Note:* The group names are based on ASTM D-2487.)

Criteria for group symbol					Criteria for group name	
F_{200}	C_u	C_c	Relationship between LL & PI	Group symbol	GF	Group name
<5	≥4	$1 \le C_c \le 3$		GW	<15	Well graded gravel
					≥15	Well graded gravel with sand
	$C_u < 4$ and/or $1 > C_c > 3$			GP	<15	Poorly graded gravel
					≥15	Poorly graded gravel with sand
>12			$PI < 4$ or $PI < 0.73(LL - 20)$	GM	<15	Silty gravel
					≥15	Silty gravel with sand
			$PI > 7$ and $PI \ge 0.73(LL - 20)$	GC	<15	Clayey gravel
					≥15	Clayey gravel with sand
			$4 \le PI \le 7$ and $PI \ge 0.73(LL - 20)$	GC-GM	<15	Silty, clayey gravel
					≥15	Silty, clayey gravel with sand
$5 \le F_{200} \le 12$	≥4	$1 \le C_c \le 3$	$PI < 4$ or $PI < 0.73(LL - 20)$	GW-GM	<15	Well graded gravel with silt
					≥15	Well graded gravel with silt and sand
			$PI > 7$ and $PI \ge 0.73(LL - 20)$	GW-GC	<15	Well graded gravel with clay
					≥15	Well graded gravel with clay and sand
	$C_u < 4$ and/or $1 > C_c > 3$		$PI < 4$ or $PI < 0.73(LL - 20)$	GP-GM	<15	Poorly graded gravel with silt
					≥15	Poorly graded gravel with silt and sand
			$PI > 7$ and $PI \ge 0.73(LL - 20)$	GP-GC	<15	Poorly graded gravel with clay
					≥15	Poorly graded gravel with clay and sand

Table 9–6. Unified Classification of Sandy Soils ($R_4 \leq 0.5R_{200}$) (*Note:* The group names are based on ASTM D-2487.)

Criteria for group symbol				Group symbol	Criteria for group name	Group name
F_{200}	C_u	C_c	Relationship between *LL* & *PI*		*GF*	
<5	≥6	$1 \leq C_c \leq 3$		*SW*	<15	Well graded sand
					≥15	Well graded sand with gravel
	$C_u < 6$ and/or $1 > C_c > 3$			*SP*	<15	Poorly graded sand
					≥15	Poorly graded sand with gravel
>12			$PI < 4$ or $PI < 0.73(LL - 20)$	*SM*	<15	Silty sand
					≥15	Silty sand with gravel
			$PI > 7$ and $PI \geq 0.73(LL - 20)$	*SC*	<15	Clayey sand
					≥15	Clayey sand with gravel
			$4 \leq PI \leq 7$ and $PI \geq 0.73(LL - 20)$	*SC-SM*	<15	Silty, clayey sand
					≥15	Silty, clayey sand with gravel
$5 \leq F_{200} \leq 12$	≥6	$1 \leq C_c \leq 3$	$PI < 4$ or $PI < 0.73(LL - 20)$	*SW-SM*	<15	Well graded sand with silt
					≥15	Well graded sand with silt and gravel
			$PI > 7$ and $PI \geq 0.73(LL - 20)$	*SW-SC*	<15	Well graded sand with clay
					≥15	Well graded sand with clay and gravel
	$C_u < 6$ and/or $1 > C_c > 3$		$PI < 4$ or $PI < 0.73(LL - 20)$	*SP-SM*	<15	Poorly graded sand with silt
					≥15	Poorly graded sand with silt and gravel
			$PI > 7$ and $PI \geq 0.73(LL - 20)$	*SP-SC*	<15	Poorly graded sand with clay
					≥15	Poorly graded sand with clay and gravel

Example 9–2

Classify Soils A and B as given in Example 9–1 and obtain the group symbols and group names. Assume Soil B to be inorganic.

Soil A:	Percent passing No. 4 sieve	=	98
	Percent passing No. 10 sieve	=	90
	Percent passing No. 40 sieve	=	76
	Percent passing No. 200 sieve	=	34
	Liquid limit	=	38
	Plastic limit	=	26

Soil B:	Percent passing No. 4 sieve	=	100
	Percent passing No. 10 sieve	=	98
	Percent passing No. 40 sieve	=	86
	Percent passing No. 200 sieve	=	58
	Liquid limit	=	49
	Plastic limit	=	28

Solution

Soil A:

Step 1. $F_{200} = 34\%$

Step 2. $R_{200} = 100 - F_{200} = 100 - 34 = 66\%$

Step 3. $R_{200} = 66\% > 50\%$. So it is a coarse-grained soil.

Skip Step 4.

Step 5. $R_4 = 100 - F_4 = 2\%$

$R_4 < 0.5R_{200} = 33\%$

So it is a sandy soil (Step 5b). $F_{200} > 12\%$. Thus C_u and C_c values are not needed.

$PI = LL - PL = 38 - 26 = 12$

$PI = 12 < 0.73(LL - 20) = 0.73(38 - 20) = 13.14$

From Table 9–6, the *group symbol* is **SM**.

$GF = R_4 = 2\%$ (which is $< 15\%$)

From Table 9–6, the *group name* is **silty sand**.

Soil B:

Step 1. $F_{200} = 58\%$

Step 2. $R_{200} = 100 - F_{200} = 100 - 58 = 42\%$

Step 3. $R_{200} = 42\% < 50\%$. So it is a fine-grained soil.

Step 4. From Table 9–3, $LL = 49 < 50$

$PI = 49 - 28 = 21$

$PI = 21 < 0.73(LL - 20) = 0.73(49 - 20) = 21.17$

So the *group symbol* is **ML**.

Again, $R_{200} = 42\% > 30\%$

$R_4 = 100 - F_4 = 100 - 100 = 0\%$

So $GF = 0\% < 15\%$

$SF = R_{200} - GF = 42 - 0 = 42\%$

$SF/GF > 1$

So the *group name* is **sandy silt**.

10

Constant Head Permeability Test in Sand

Introduction

The rate of flow of water through a soil specimen of gross cross-sectional area, A, can be expressed as

$$q = kiA \tag{10.1}$$

where q = flow in unit time
k = coefficient of permeability
i = hydraulic gradient

For coarse sands, the value of the coefficient of permeability may vary from 1 to 0.01 cm/s and, for fine sand, it may be in the range of 0.01 to 0.001 cm/s.

Several relations between k and the void ratio, e, for sandy soils have been proposed. They are of the form

$$k \propto e^2 \tag{10.2}$$

$$k \propto \frac{e^2}{1+e} \tag{10.3}$$

$$k \propto \frac{e^3}{1+e} \tag{10.4}$$

The coefficient of permeability of sands can be easily determined in the laboratory by two simple methods. They are (a) the constant head test and (b) the variable head test. In this chapter, the *constant head test method* will be discussed.

Equipment

1. Constant head permeameter
2. Graduated cylinder (250 cc or 500 cc)
3. Balance, sensitive up to 0.1 g
4. Thermometer, sensitive up to 0.1°C
5. Rubber tubing
6. Stop watch

Constant Head Permeameter

A schematic diagram of a constant head permeameter is shown in Fig. 10–1. This can be assembled in the laboratory at very low cost. It essentially consists of a plastic soil specimen cylinder, two porous stones, two rubber stoppers, one spring, one constant head chamber, a large funnel, a stand, a scale, three clamps, and some plastic tubes. The plastic cylinder may have an inside diameter of 2.5 in. (63.5 mm). This is because 2.5 in. (63.5 mm) diameter porous stones are usually available in most soils laboratories. The length of the specimen tube may be about 12 in. (304.8 mm).

Procedure

1. Determine the mass of the plastic specimen tube, the porous stones, the spring, and the two rubber stoppers (W_1).
2. Slip the bottom porous stone into the specimen tube, and then fix the bottom rubber stopper to the specimen tube.
3. Collect oven-dry sand in a container. Use a spoon, pour the sand into the specimen tube in small layers, and compact it by vibration and/or other compacting means.
   Note: By changing the degree of compaction, a number of test specimens having different void ratios can be prepared.
4. When the length of the specimen tube is about two-third the length of the tube, slip the top porous stone into the tube to rest firmly on the specimen.
5. Place a spring on the top porous stone, if necessary.
6. Fix a rubber stopper to the top of the specimen tube.
   Note: The spring in the assembled position will not allow any expansion of the specimen volume, and thus the void ratio, during the test.
7. Determine the mass of the assembly (Step 6 – W_2).
8. Measure the length (L) of the compacted specimen in the tube.
9. Assemble the permeameter near a sink, as shown in Fig. 10–1.
10. Run water into the top of the large funnel fixed to the stand through a plastic tube from the water inlet. The water will flow through the specimen to the constant head chamber. After some time, the water will flow into the sink through the outlet in the constant head chamber.

Note: Make sure that water does not leak from the specimen tube.

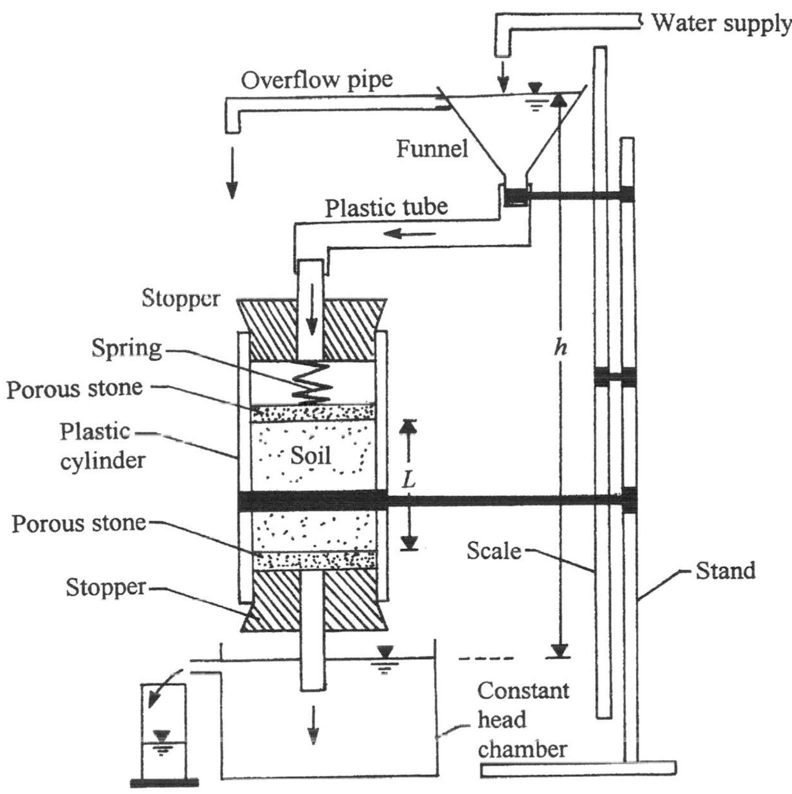

Figure 10–1. Schematic diagram of constant head permeability test setup.

11. Adjust the supply of water to the funnel so that the water level in the funnel remains constant. At the same time, allow the flow to continue for about 10 minutes in order to saturate the specimen.

    Note: Some air bubbles may appear in the plastic tube connecting the funnel to the specimen tube. Remove the air bubbles.
12. After a steady flow is established (that is, once the head difference h is constant), collect the water flowing out of the constant head chamber (Q) in a graduated cylinder. Record the collection time (t) with a stop watch.
13. Repeat Step 12 three times. Keep the collection time (t) the same and determine Q. Then find the average value of Q.
14. Change the head difference, h, and repeat Steps 11, 12 and 13 about three times.
15. Record the temperature, T, of the water to the nearest degree.

    Note: This value is sufficiently accurate for this type of test.

Calculation

1. Calculate the void ratio of the compacted specimen as follows:
   Dry density, ρ_d, of the soil specimen as

$$\rho_d = \frac{W_2 - W_1}{\frac{\pi}{4} D^2 L}$$

   Thus

$$e = \frac{G_s \rho_w}{\rho_d} - 1 \tag{10.5}$$

   where G_s = specific gravity of soil solids
   ρ_w = density of water
   D = diameter of the specimen
   L = length of the specimen

2. Calculate k as

$$k = \frac{QL}{Aht} \tag{10.6}$$

   where A = area of specimen = $\frac{\pi}{4} D^2$

3. The value k is usually given for a test temperature of water at 20°C. So calculate $k_{20°C}$ as

$$k_{20°C} = k_{T°C} \frac{\eta_{T°C}}{\eta_{20°C}} \tag{10.7}$$

   where $\eta_{T°C}$ and $\eta_{20°C}$ are viscosities of water at T°C and 20°C, respectively.

   Table 10–1 gives the values of $\frac{\eta_{T°C}}{\eta_{20°C}}$ for various values of T (in °C).

   Tables 10–2 and 10–3 give sample calculations for the permeability test.

Table 10–1. Variation of $\eta_{T°C}/\eta_{20°C}$

Temperature, T (°C)	$\eta_{T°C}/\eta_{20°C}$	Temperature, T (°C)	$\eta_{T°C}/\eta_{20°C}$
15	1.135	23	0.931
16	1.106	24	0.910
17	1.077	25	0.889
18	1.051	26	0.869
19	1.025	27	0.850
20	1.000	28	0.832
21	0.976	29	0.814
22	0.953	30	0.797

Table 10–2. Constant Head Permeability Test

Determination of Void Ratio of Specimen

Description of soil *Uniform sand* Sample No. ______

Location ______

Length of specimen, L *13.2* cm Diameter of specimen, D *6.35* cm

Tested by ______ Date ______

Volume of specimen, $V = \frac{\pi}{4} D^2 L$ (cm^2)	*418.03*
Specific gravity of soil solids, G_s	*2.66*
Mass of specimen tube with fittings, W_1 (g)	*238.4*
Mass of tube with fittings and specimen, W_2 (g)	*965.3*
Dry density of specimen, $\rho_d = \frac{W_2 - W_1}{V}$ (g / cm^3)	*1.74*
Void ratio of specimen, $e = \frac{G_s \rho_w}{\rho_d} - 1$ (*Note:* $\rho_w = 1$ g/cm^3)	*0.53*

Table 10–3. Constant Head Permeability Test

Determination of Coefficient of Permeability

Test No.	1	2	3
Average flow, Q (cm^3)	305	375	395
Time of collection, t (s)	60	60	60
Temperature of water, T (°C)	25	25	25
Head difference, h (cm)	60	70	80
Diameter of specimen, D (cm)	6.35	6.35	6.35
Length of specimen, L (cm)	13.2	13.2	13.2
Area of specimen, $A = \frac{\pi}{4} D^2$ (cm^2)	31.67	31.67	31.67
$k = \frac{QL}{Aht}$ (cm / s)	0.035	0.037	0.034
Average k = 0.035 cm/s			
$k_{20°C} = k_{T°C} \frac{\eta_{T°C}}{\eta_{20°C}}$ = 0.035(0.889) = 0.031 cm/s			

11
Falling Head Permeability Test in Sand

Introduction

The procedure for conducting the constant head permeability tests in sand were discussed in the preceding chapter. The falling head permeability test is another experimental procedure to determine the coefficient of permeability of sand.

Equipment

1. Falling head permeameter
2. Balance sensitive to 0.1 g
3. Thermometer
4. Stop watch

Falling Head Permeameter

A schematic diagram of a falling head permeameter is shown in Fig. 11–1. This consists of a specimen tube essentially the same as that used in the constant head test. The top of the specimen tube is connected to a burette by plastic tubing. The specimen tube and the burette are held vertically by clamps from a stand. The bottom of the specimen tube is connected to a plastic funnel by a plastic tube. The funnel is held vertically by a clamp from another stand. A scale is also fixed vertically to this stand.

Procedure

Steps 1 through 9: Follow the same procedure (i.e., Steps 1 through 9) as described in Chapter 10 for the preparation of the specimen.

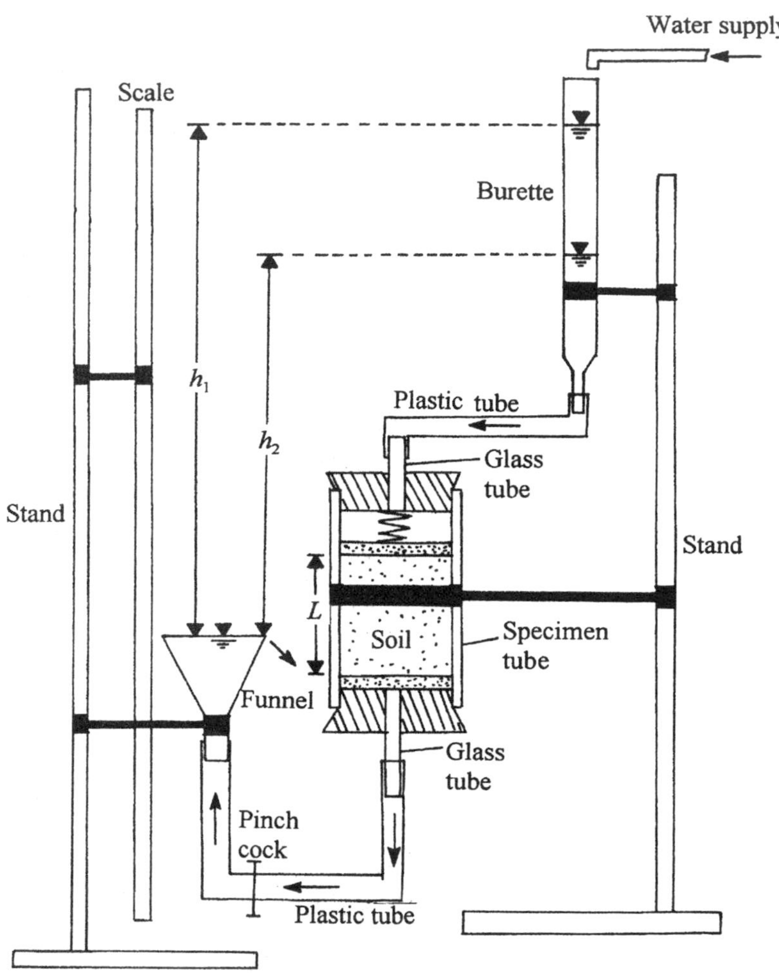

Figure 11–1. Schematic diagram of falling head permeability test setup.

10. Supply water using a plastic tube from the water inlet to the burette. The water will flow from the burette to the specimen and then to the funnel. Check to see that there is no leak. Remove all air bubbles.
11. Allow the water to flow for some time in order to saturate the specimen. When the funnel is full, water will flow out of it into the sink.
12. Using the pinch cock, close the flow of water through the specimen. The pinch cock is located on the plastic pipe connecting the bottom of the specimen to the funnel.
13. Measure the head difference, h_1 (cm) (see Fig. 11–1).
    Note: Do not add any more water to the burette.
14. Open the pinch cock. Water will flow through the burette to the specimen and then out of the funnel. Record time (t) with a stop watch until the head difference is equal to h_2 (cm) (Fig. 11–1). Close the flow of water through the specimen using the pinch cock.

15. Determine the volume (V_w) of water that is drained from burette in cm^3.
16. Add more water to the burette to make another run. Repeat Steps 13, 14 and 15. However, h_1 and h_2 should be changed for each run.
17. Record the temperature, T, of the water to the nearest degree (°C).

Calculation

The coefficient of permeability can be expressed by the relation

$$k = 2.303 \frac{aL}{At} \log \frac{h_1}{h_2} \tag{11.1}$$

where a = inside cross-sectional area of the burette

[For an example for derivation, see Das (1994) under "References" at the back of the book.]

$$a = \frac{V_w}{(h_1 - h_2)} \tag{11.2}$$

Therefore

$$k = \frac{2.303 V_w L}{(h_1 - h_2) t A} \log \frac{h_1}{h_2} \tag{11.3}$$

where A = area of the specimen

As in Chapter 10

$$k_{20°C} = k_{T°C} \frac{\eta_{T°C}}{\eta_{20°C}} \tag{11.4}$$

Sample calculations are shown in Tables 11–1 and 11–2.

Table 11–1. Falling Head Permeability Test

Determination of Void Ratio of Specimen

Description of soil *Uniform sand* Sample No. ____________

Location ____

Length of specimen, L *13.2* cm Diameter of specimen, D *6.35* cm

Tested by ____________________________ Date ____________

Volume of specimen, $V = \frac{\pi}{4} D^2 L \ (\text{cm}^2)$	418.03
Specific gravity of soil solids, G_s	2.66
Mass of specimen tube with fittings, W_1 (g)	238.4
Mass of tube with fittings and specimen, W_2 (g)	965.3
Dry density of specimen, $\rho_d = \frac{W_2 - W_1}{V} \ (\text{g/cm}^3)$	1.74
Void ratio of specimen, $e = \frac{G_s \rho_w}{\rho_d} - 1$ (*Note:* $\rho_w =$	0.53

Table 11–2. Falling Head Permeability Test
Determination of Coefficient of Permeability

Test No.	1	2	3
Diameter of specimen, D (cm)	6.35	6.35	6.35
Length of specimen, L (cm)	13.2	13.2	13.2
Area of specimen, A (cm^2)	31.67	31.67	31.67
Beginning head difference, h_1 (cm)	85.0	76.0	65.0
Ending head difference, h_2 (cm)	24.0	20.0	20.0
Test duration, t (s)	15.4	15.3	14.4
Volume of water flow through the specimen, V_w (cm^3)	64	58	47
$k = \dfrac{2.303 V_w L}{(h_1 - h_2) t A} \log \dfrac{h_1}{h_2}$ (cm / s^2)	0.036	0.038	0.036
Average k = 0.037 cm/s			
$k_{20^\circ C} = k_{T^\circ C} \dfrac{\eta_{T^\circ C}}{\eta_{20^\circ C}}$ = (0.037)(0.889) = 0.033 cm/s			

12
Standard Proctor Compaction Test

Introduction

For construction of highways, airports, and other structures, it is often necessary to compact soil to improve its strength. Proctor (1933) developed a laboratory compaction test procedure to determine the maximum dry unit weight of compaction of soils which can be used for specification of field compaction. This test is referred to as the *standard Proctor compaction test* and is based on the compaction of the soil fraction passing No. 4 U.S. sieve.

Equipment

1. Compaction mold
2. No. 4 U.S. sieve
3. Standard Proctor hammer (5.5 lb)
4. Balance sensitive up to 0.01 lb
5. Balance sensitive up to 0.1 g
6. Large flat pan
7. Jack
8. Steel straight edge
9. Moisture cans
10. Drying oven
11. Plastic squeeze bottle with water

Figure 12–1 shows the equipment required for the compaction test with the exception of the jack, the balances, and the oven.

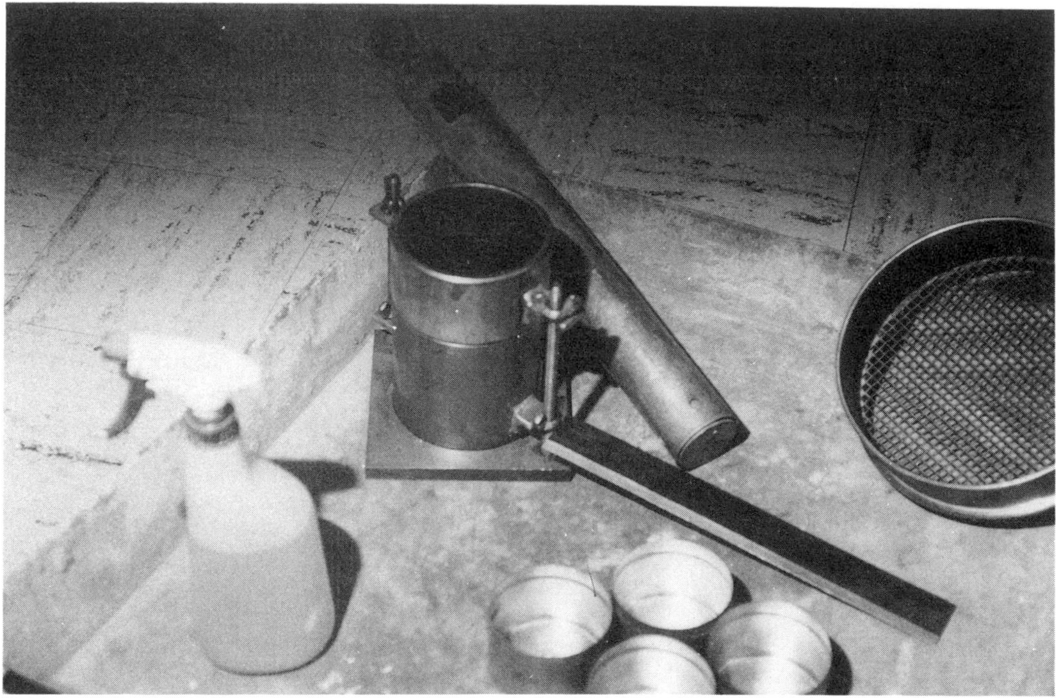

Figure 12–1. Equipment for Proctor compaction test.

Proctor Compaction Mold and Hammer

A schematic diagram of the Proctor compaction mold, which is 4 in. (101.6 mm) in diameter and 4.584 in. (116.4) in height, is shown in Fig. 12–2*a*. There is a base plate and an extension that can be attached to the top and bottom of the mold, respectively. The inside of the mold is $1/30$ ft^3 (943.9 cm^3).

Figure 12–2*b* shows the schematic diagram of a standard Proctor hammer. The hammer can be lifted and dropped through a vertical distance of 12 in. (304.8 mm).

Procedure

1. Obtain about 10 lb (4.5 kg) of air-dry soil on which the compaction test is to be conducted. Break all the soil lumps.
2. Sieve the soil on a No. 4 U.S. sieve. Collect all of the minus–4 material in a large pan. This should be about 6 lb (2.7 kg) or more.
3. Add enough water to the minus–4 material and mix it in thoroughly to bring the moisture content up to about 5%.
4. Determine the weight of the Proctor mold + base plate (not the extension), W_1, (lb).
5. Now attach the extension to the top of the mold.
6. Pour the moist soil into the mold in *three* equal layers. Each layer should be compacted uniformly by the standard Proctor hammer *25 times* before the next layer of loose soil is poured into the mold.

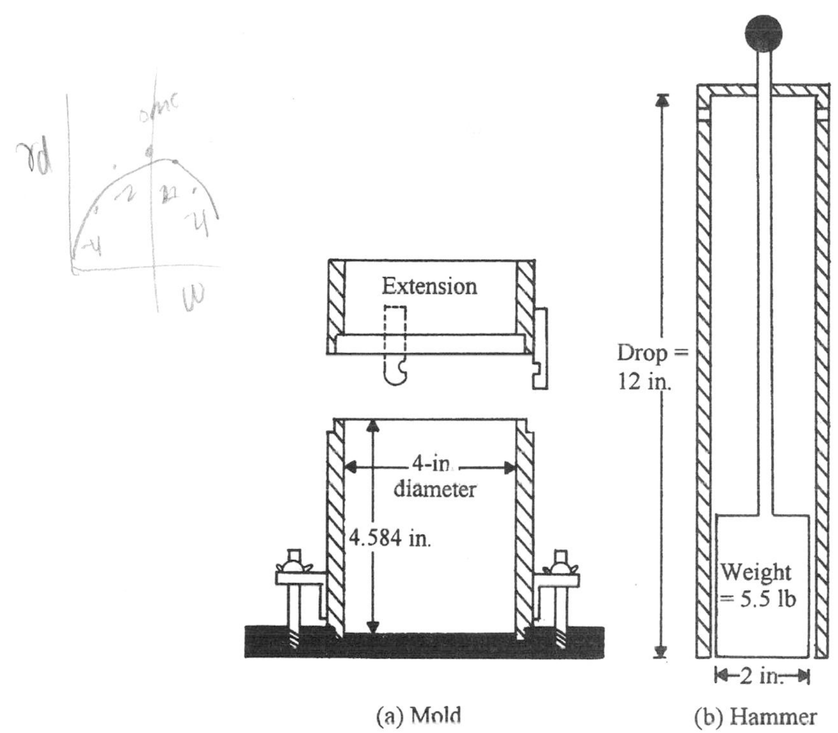

Figure 12–2. Standard Proctor mold and hammer.

Note: The layers of loose soil that are being poured into the mold should be such that, *at the end of the three-layer compaction*, the soil should extend *slightly above* the top of the rim of the compaction mold.

7. Remove the top attachment from the mold. Be careful not to break off any of the compacted soil inside the mold while removing the top attachment.
8. Using a straight edge, trim the excess soil above the mold (Fig. 12–3). Now the top of the compacted soil will be even with the top of the mold.
9. Determine the weight of the mold + base plate + compacted moist soil in the mold, W_2 (lb).
10. Remove the base plate from the mold. Using a jack, extrude the compacted soil cylinder from the mold.
11. Take a moisture can and determine its mass, W_3 (g).
12. From the moist soil extruded in Step 10, collect a moisture sample in the moisture can (Step 11) and determine the mass of the can + moist soil, W_4 (g).
13. Place the moisture can with the moist soil in the oven to dry to a constant weight.
14. Break the rest of the compacted soil (to No. 4 size) by hand and mix it with the left-over moist soil in the pan. Add more water and mix it to raise the moisture content by about 2%.

Figure 12–3. Excess soil being trimmed (Step 8).

15. Repeat Steps 6 through 12. In this process, the weight of the mold + base plate + moist soil (W_2) will first increase with the increase in moisture content and then decrease. Continue the test until at least two successive down readings are obtained.
16. The next day, determine the mass of the moisture cans + soil samples, W_5 (g) (from Step 13).

Calculation

Dry Unit Weight and Moisture Content at Compaction

The sample calculations for a standard Proctor compaction test are given in Table 12–1. Referring to Table 12–1,

Line 1 — Weight of mold, W_1, to be determined from test (Step 4).

Line 2 — Weight of mold + moist compacted soil, W_2, to be determined from test (Step 9).

Line 3 — Weight of moist compacted soil = $W_2 - W_1$ (Line 2 - Line 1).

Line 4 — Moist unit weight

$$\gamma = \frac{\text{weight of compacted moist soil}}{\text{volume of mold}} = \frac{W_2 - W_1 \text{ (lb)}}{1/30 \text{ ft}^3}$$

$$= (30 \text{ lb}/\text{ft}^3) \times (\text{Line 3})$$

Line 6 — Mass of moisture can, W_3, to be determined from test (Step 11).

Line 7 — Mass of moisture can + moist soil, W_4, to be determined from test (Step 12).
Line 8 — Mass of moisture can + dry soil, W_5, to be determined from test (Step 16).
Line 9 — Compaction moisture content

$$w\ (\%) = \frac{W_4 - W_5}{W_5 - W_3} \times 100$$

Line 10 — Dry unit weight

$$\gamma_d = \frac{\gamma}{1 + \dfrac{w\ (\%)}{100}} = \frac{\text{Line 4}}{1 + \dfrac{\text{Line 9}}{100}}$$

Zero-Air-Void Unit Weight

The maximum theoretical dry unit weight of a compacted soil at a given moisture content will occur when there is no air left in the void spaces of the compacted soil. This can be given by

$$\gamma_{d(\text{theory-max})} = \gamma_{zav} = \frac{\gamma_w}{\dfrac{w\ (\%)}{100} + \dfrac{1}{G_s}} \tag{12.1}$$

where γ_{zav} = zero-air-void unit weight
γ_w = unit weight of water
w = moisture content
G_s = specific gravity of soil solids.

Since the values of γ_w and G_s will be known, several values of w (%) can be assumed and γ_{zav} can be calculated. Table 12–2 shows the calculations for γ_{zav} for the soil tested and reported in Table 12–1.

Graph

Plot a graph showing γ_d (Line 10, Table 12–1) versus w (%) (Line 9, Table 12–1) and determine the *maximum dry unit weight of compaction* [$\gamma_{d(\max)}$]. Also determine the *optimum moisture content,* w_{opt}, which is the moisture content corresponding to $\gamma_{d(\max)}$. On the same graph, plot γ_{zav} versus w (%).

Note: For a given soil, *no portion* of the experiment curve of γ_d versus w (%) should plot to the *right* of the zero-air-void curve.

Figure 12–4 shows the results of calculations made in Tables 12–1 and 12–2.

Table 12–1. Standard Proctor Compaction Test
Determination of Dry Unit Weight

Description of soil *Light brown clayey silt* Sample No. *2*

Location ______

Volume of mold *1/30* ft³ Weight of hammer *5.5* lb Number of blows/layer *25* Number of layers *3*

Tested by ______ Date ______

Test	1	2	3	4	5	6
1. Weight of mold, W_1 (lb)	10.35	10.35	10.35	10.35	10.35	10.35
2. Weight of mold + moist soil, W_2 (lb)	14.19	14.41	14.53	14.63	14.51	14.47
3. Weight of moist soil, $W_2 - W_1$ (lb)	3.84	4.06	4.18	4.28	4.16	4.12
4. Moist unit weight, $\gamma = \frac{W_2 - W_1}{1/30}$ (lb / ft³)	115.2	121.8	125.4	128.4	124.8	123.8
5. Moisture can number	202	212	222	242	206	504
6. Mass of moisture can, W_3 (g)	54.0	53.3	53.3	54.0	54.8	40.8
7. Mass of can + moist soil, W_4 (g)	253.0	354.0	439.0	490.0	422.8	243.0
8. Mass of can + dry soil, W_5 (g)	237.0	326.0	401.0	441.5	374.7	211.1
9. Moisture content, $w\,(\%) = \frac{W_4 - W_5}{W_5 - W_3} \times 100$	8.7	10.3	10.9	12.5	15.0	18.8
10. Dry unit weight of compaction γ_d (lb / ft³) $= \frac{\gamma}{1 + \frac{w\,(\%)}{100}}$	106.0	110.4	113.0	114.1	108.5	104.2

Table 12–2. Standard Proctor Compaction Test
Zero-Air-Void Unit Weight

Description of soil Light brown clayey silt Sample No. 2

Location

Tested by Date

Specific gravity of soil solids, G_s	Assumed moisture content, w (%)	Unit weight of water, γ_w (lb/ft³)	γ_{zav}[a] (lb/ft³)
2.68	10	62.4	131.9
2.68	12	62.4	126.5
2.68	14	62.4	121.6
2.68	16	62.4	117.0
2.68	18	62.4	112.8
2.68	20	62.4	108.7

[a] Eq. (12.1)

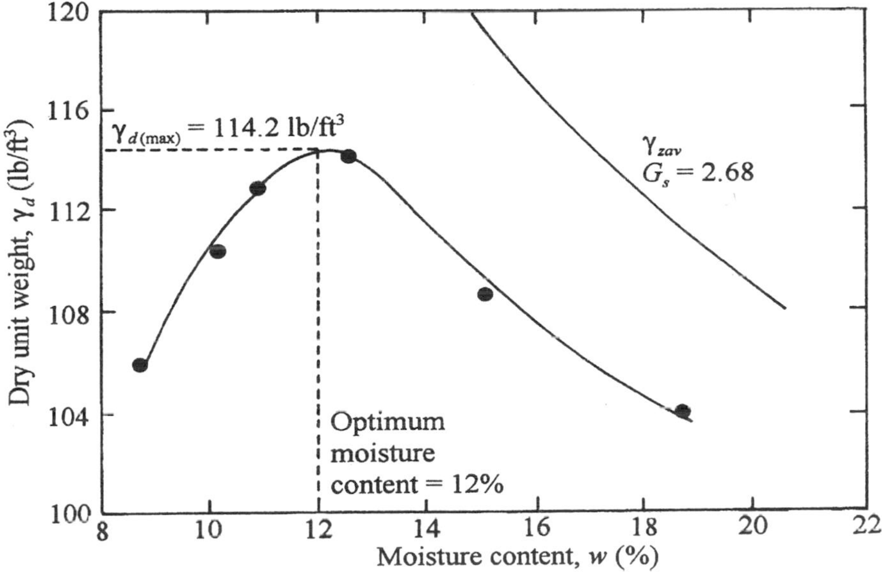

Figure 12–4. Plot of γ_d vs. w (%) and γ_{zav} vs. w (%) for test results reported in Tables 12–1 and 12–2.

General Comments

In most of the specifications for earth work, it is required to achieve a compacted field dry unit weight of 90% to 95% of the maximum dry unit weight obtained in the laboratory. This is sometimes referred to as relative compaction, R, or

$$R\,(\%) = \frac{\gamma_{d\,(\text{field})}}{\gamma_{d\,(\text{max - lab})}} \times 100 \tag{12.2}$$

For granular soils, it can be shown that

$$R\,(\%) = \frac{R_o}{1 - D_r\,(1 - R_o)} \times 100 \tag{12.3}$$

where D_r = relative density of compaction

$$R_o = \frac{\gamma_{d\,(\text{max})}}{\gamma_{d\,(\text{min})}} \tag{12.4}$$

Compaction of cohesive soils will influence its structure, coefficient of permeability, one-dimensional compressibility and strength. For further discussion on this topic, refer to Das (1994).

In this chapter, the laboratory test outlines given for compaction tests use the following:

Volume of mold = $1/30$ ft^3

Number of blows = 25

These values are generally used for fine-grained soils that pass through No. 4 U.S. sieve. However, ASTM and AASHTO have four different methods for the standard Proctor compaction test that reflect the size of the mold, the number of blows per layer, and the maximum particle size in a soil used for testing. Summaries of these methods are given in Table 12–3.

Table 12–3. Summary of Standard Proctor Compaction Test Specifications (ASTM D-698, AASHTO T-99)

Description	Method A	Method B	Method C	Method D
Mold:				
Volume (ft^3)	1/30	1/13.33	1/30	1/13.33
Height (in.)	4.58	4.58	4.58	4.58
Diameter (in.)	4	6	4	6
Weight of hammer (lb)	5.5	5.5	5.5	5.5
Height of drop of hammer (in.)	12	12	12	12
Number of layers of soil	3	3	3	3
Number of blows per layer	25	56	25	56
Test on soil fraction passing sieve	No. 4	No. 4	¾ in.	¾ in.

13 Modified Proctor Compaction Test

Introduction

In the preceding chapter, we have seen that water generally acts as a lubricant between solid particles during the soil compaction process. Because of this, in the initial stages of compaction, the dry unit weight of compaction increases. However another factor that will control the dry unit weight of compaction of a soil at a given moisture content is the energy of com-paction. For the standard Proctor compaction test, the energy of compaction can be given by

$$\frac{(3\text{ layers})(25\text{ blows / layer})(5.5\text{ lb})(1\text{ ft / blow})}{\frac{1}{30}\text{ ft}^3} = 12{,}375\frac{\text{ft}\cdot\text{lb}}{\text{ft}^3}\ (593\text{ kJ / m}^3)$$

The modified Proctor compaction test is a standard test procedure for compaction of soil using a higher energy of compaction. In this test, the compaction energy is equal to

$$56{,}250\frac{\text{ft}\cdot\text{lb}}{\text{ft}^3}\ (2694\text{ kJ / m}^3)$$

Equipment

The equipment required for the modified Proctor compaction test is the same as in Chapter 12 with the exception of the standard Proctor hammer (Item 3). The hammer used for this test weighs 10 lb and drops through a vertical distance of 18 in. Figure 13–1 shows the standard and modified Proctor test hammers side by side.

The compaction mold used in this test is the same as described in Chapter 12 (i.e., volume = $\frac{1}{30}$ ft^3.

Figure 13-1. Comparison of the standard and modified Proctor compaction hammer. *Note*: The left-side hammer is for the modified Proctor compaction test.

Procedure

The procedure is the same as described in Chapter 12, except for Item 6. The moist soil has to be poured into the mold in five equal layers. Each layer has to be compacted by the modified Proctor hammer with 25 blows per layer.

Calculation, Graph, and Zero-Air-Void Curve

Same as in Chapter 12.

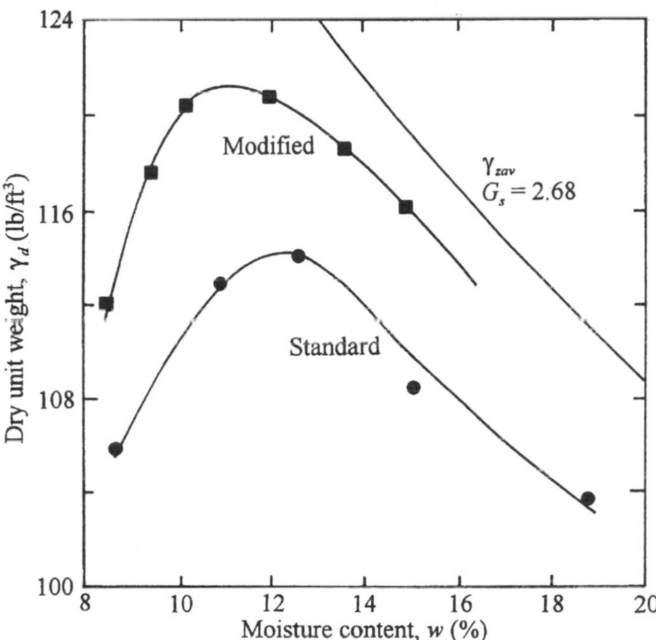

Figure 13–2. Comparison of standard and modified Proctor compaction test results for the soil reported in Tables 12–1 and 12–2.

General Comments

1. The modified Proctor compaction test results for the same soil as reported in Tables 12–1 and 12–2 and Fig. 12–4 are shown in Fig. 13–2. A comparison of γ_d vs. w (%) curves obtained from standard and modified Proctor compaction tests shows that
   (a) The maximum dry unit weight of compaction increases with the increase in the compacting energy, and
   (b) The optimum moisture content decreases with the increase in the energy of compaction
2. As reported in Chapter 12, there are four different methods suggested by ASTM and AASHTO for this test, and they are shown in Table 13–1.

Table 13–1. Summary of Modified Proctor Compaction Test Specifications (ASTM D-1557, AASHTOT-180)

Description	Method A	Method B	Method C	Method D
Mold:				
Volume (ft^3)	$^1/_{30}$	$^1/_{13.33}$	$^1/_{30}$	$^1/_{13.33}$
Height (in.)	4.58	4.58	4.58	4.58
Diameter (in.)	4	6	4	6
Weight of hammer (lb)	10	10	10	10
Height of drop of hammer (in.)	18	18	18	18
Number of layers of soil	5	5	5	5
Number of blows per layer	25	56	25	56
Test on soil fraction passing sieve	No. 4	No. 4	¾ in.	¾ in.

14
Determination of Field Unit Weight of Compaction by Sand Cone Method

Introduction

In the field during soil compaction, it is sometimes necessary to check the compacted dry unit weight of soil and compare it with the specifications drawn up for the construction. One of the simplest methods of determining the field unit weight of compaction is by the sand cone method, which will be described in this chapter.

Equipment

1. Sand cone apparatus which consists of a one-gallon glass or plastic bottle with a metal cone attached to it.
2. Base plate
3. One-gallon can with cap
4. Tools to dig a small hole in the field
5. Balance with minimum readability of 0.01 lb
6. 20-30 Ottawa sand
7. Proctor compaction mold without attached extension
8. Steel straight edge

Figure 14–1 shows the assembly of the equipment necessary for determination of the field unit weight.

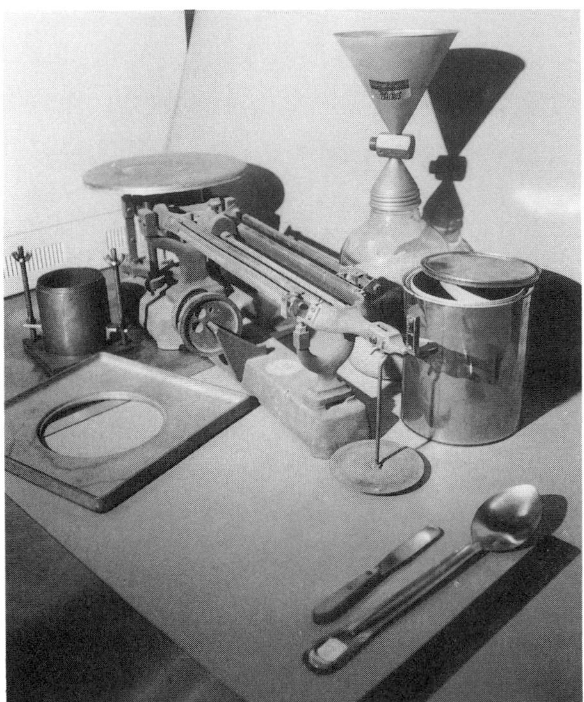

Figure 14–1. Assembly of equipment necessary for determination of field unit weight of compaction.

Procedure – Laboratory Work

Determination of Dry Unit Weight of 20-30 Ottawa Sand

1. Determine the weight of the Proctor compaction mold, W_1.
2. Using a spoon, fill the compaction mold with 20-30 Ottawa sand. Avoid any vibration or other means of compaction of the sand poured into the mold. When the mold is full, strike off the top of the mold with the steel straight edge. Determine the weight of the mold and sand, W_2.

Calibration of Cone

3. We need to determine the weight of the Ottawa sand that is required to fill the cone. This can be done by filling the one-gallon bottle with Ottawa sand. Determine the weight of the bottle + cone + sand, W_3. Close the valve of the cone which is attached to the bottle. Place the base plate on a flat surface. Turn the bottle with the cone attached to it upside down and place the open mouth of the cone in the center hole of the base plate (Figure 14–2). Open the cone valve. Sand will flow out of the bottle and gradually fill the cone. When the cone is filled with sand, the flow of sand from the bottle will stop. Close the cone valve. Remove the bottle and cone combination from the base plate and determine its weight, W_4.

Preparation for Field Work

4. Determine the weight of the gallon can without the cap, W_5.

Figure 14–2. Calibration of the sand cone.

5. Fill the one-gallon bottle (with the sand cone attached to it) with sand. Close the valve of the cone. Determine the weight of the bottle + cone + sand, W_6.

Procedure – Field Work

6. Now proceed to the field with the bottle with the cone attached to it (filled with Ottawa sand—Step 4), the base plate, the digging tools, and the one-gallon can with its cap.
7, Place the base plate on a level ground in the field. Under the center hole of the base plate, dig a hole in the ground using the digging tools. The volume of the hole should be smaller than the volume of the sand in the bottle minus the volume of the cone.
8. Remove *all* the loose soil from the hole and put it in the gallon can. Close the cap tightly so as not to lose any moisture. Be careful not to move the base plate.
9. Turn the gallon bottle filled with sand with cone attached to it upside down and place it on the center of the base plate. Open the valve of the cone. Sand will flow from the bottle to fill the hole in the ground and the cone. When the flow of sand from the bottle stops, close the valve of the cone and remove it.
10. Bring all the equipment back to the laboratory. Determine the weight of the gallon can + moist soil from the field (without the cap), W_7. Also determine the weight of the bottle + can + sand after use, W_8.

11. Put the gallon can with the moist soil in the oven to dry to a constant weight. Determine the weight of the can without the cap + oven-dry soil, W_9.

Calculation

A sample calculation for determination of dry unit weight of field compaction by the sand cone method is given in Table 14–1. With reference to the table, following are the required calculations.

1. Calculate the dry unit weight of sand (Line 4)

$$\gamma_{d\,(\text{sand})} = \frac{W_2 - W_1}{V_1} = \frac{\text{Line 2} - \text{Line 1}}{\text{Line 3}} \tag{14.1}$$

where V_1 = volume of Proctor mold

2. Calculate the weight of sand to fill the cone (Line 7)

$$W_c = W_4 - W_3 = \text{Line 6} - \text{Line 5} \tag{14.2}$$

3. Calculate the volume of the hole in the field (Line 10)

$$V_2 = \frac{W_6 - W_8 - W_c}{\gamma_{d\,(\text{sand})}} = \frac{\text{Line 8} - \text{Line 9} - \text{Line 7}}{\text{Line 4}} \tag{14.3}$$

4. Calculate the moist field unit weight (Line 14)

$$\gamma = \frac{W_7 - W_5}{V_2} - \frac{\text{Line 12} - \text{Line 11}}{\text{Line 10}} \tag{14.4}$$

5. Calculate the moisture content in the field (Line 15)

$$w\,(\%) = \frac{W_7 - W_9}{W_9 - W_5} \times 100 = \frac{\text{Line 12} - \text{Line 13}}{\text{Line 13} - \text{Line 11}} \times 100 \tag{14.5}$$

6. Calculate the dry unit weight in the field (Line 16)

$$\gamma_d = \frac{\gamma}{1 + \dfrac{w\,(\%)}{100}} = \frac{\text{Line 14}}{1 + \dfrac{\text{Line 15}}{100}} \tag{14.6}$$

Table 14–1. Field Unit Weight–Sand Cone Method

Item	Quantity
Calibration of Unit Weight of Ottawa Sand	
1. Weight of Proctor mold, W_1	*10.35 lb*
2. Weight of Proctor mold + sand, W_2	*13.66 lb*
3. Volume of mold, V_1	*1/30 ft³*
4. Dry unit weight, $\gamma_{d\,(\text{sand})} = \dfrac{W_2 - W_1}{V_1}$	*99.3 lb/ft³*
Calibration Cone	
5. Weight of bottle + cone + sand (before use), W_3	*15.17 lb*
6. Weight of bottle + cone + sand (after use), W_4	*14.09 lb*
7. Weight of sand to fill the cone, $W_c = W_4 - W_3$	*1.08 lb*
Results from Field Tests	
8. Weight of bottle + cone + sand (before use), W_6	*15.42 lb*
9. Weight of bottle + cone + sand (after use), W_8	*11.74 lb*
10. Volume of hole, $V_2 = \dfrac{W_6 - W_8 - W_c}{\gamma_{d\,(\text{sand})}}$	*0.0262 ft³*
11. Weight of gallon can, W_5	*0.82 lb*
12. Weight of gallon can + moist soil, W_7	*3.92 lb*
13. Weight of gallon can + dry soil, W_9	*3.65 lb*
14. Moist unit weight of soil in field, $\gamma = \dfrac{W_7 - W_5}{V_2}$	*118.32 lb/ft³*
15. Moisture content in the field, $w\,(\%) = \dfrac{W_7 - W_9}{W_9 - W_5} \times 100$	*9.54%*
16. Dry unit weight in the field, $\gamma_{d\,(\text{sand})} = \dfrac{\gamma}{1 + \dfrac{w\,(\%)}{100}}$	*108.11 lb/ft³*

General Comments

There are at least two other methods to determine the field unit weight of compaction. They are the *rubber balloon method* (ASTM D-2167) and the *use of the nuclear density meter*. The procedure for the rubber balloon method is similar to the sand cone method, in that a test hole is made and the moist weight of the soil removed from the hole and its moisture content are determined. However, the volume of the hole is determined by introducing it into a rubber balloon filled with water from a calibrated vessel, from which the volume can be directly read.

Nuclear density meters are now used in some large projects to determine the compacted dry unit weight of a soil. The density meters operate either in drilled holes or from the ground surface. The instrument measures the weight of the wet soil per unit volume and also the weight of water present in a volume of soil. The dry unit weight of compacted soil can be determined by subtracting the weight of the water from the moist unit weight of soil.

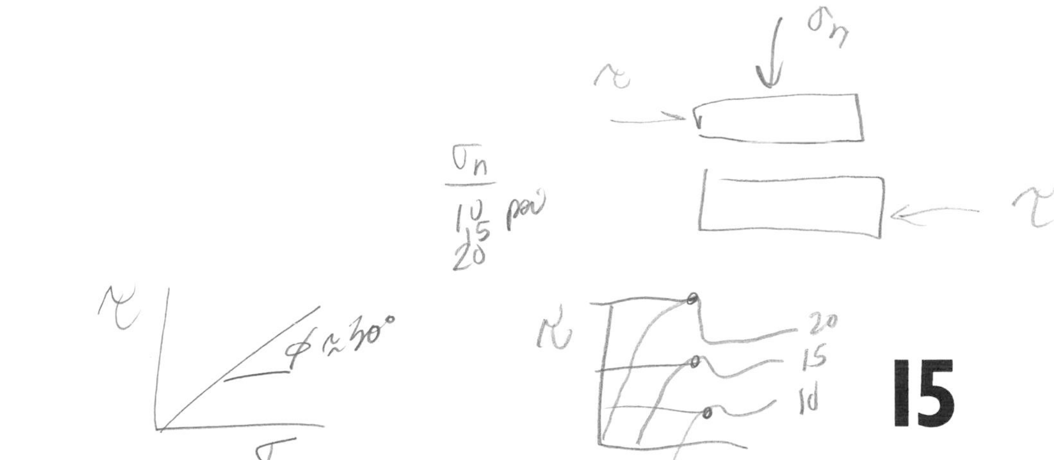

15

Direct Shear Test on Sand

Introduction

The shear strength, s, of a granular soil may be expressed by the equation

$$s = \sigma' \tan \phi \tag{15.1}$$

where σ' = effective normal stress
ϕ = angle of friction of soil

The angle of friction, ϕ, is a function of the relative density of compaction of sand, grain size, shape and distribution in a given soil mass. For a given sand, an increase in the void ratio (i.e., a decrease in the relative density of compaction) will result in a decrease of the magnitude of ϕ. However, for a given void ratio, an increase in the angularity of the soil particles will give a higher value of the soil friction angle. The general range of the angle of friction of sand with relative density is shown in Fig. 15–1.

Equipment

1. Direct shear test machine (strain controlled)
2. Balance sensitive to 0.1 g
3. Large porcelain evaporating dish
4. Tamper (for compacting sand in the direct shear box)
5. Spoon

Figure 15–2 shows a direct shear test machine. It consists primarily of a direct shear box, which is split into two halves (top and bottom) and which holds the soil specimen; a proving ring to measure the horizontal load applied to the specimen; two dial gauges (one horizontal and one vertical) to measure the deformation of the soil during the test; and a yoke by which a vertical load can be applied to the soil specimen. A horizontal load to the top half of the shear box is applied by a motor and gear arrangement. In a strain-controlled unit, the rate of movement of the top half of the shear box can be controlled.

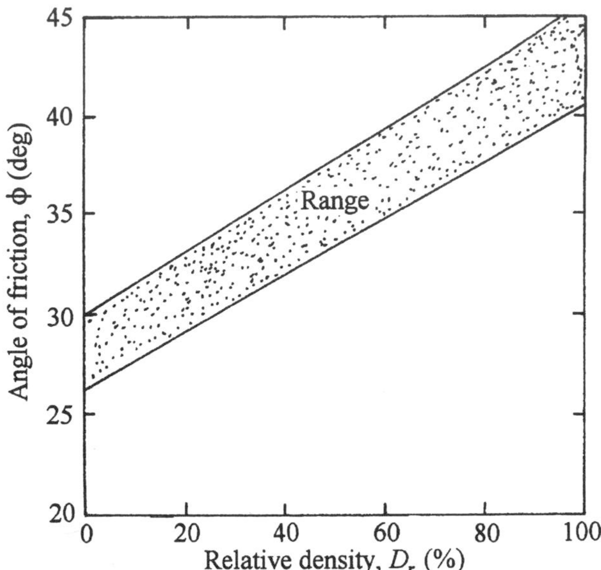

Figure 15–1. General range of the variation of angle of friction of sand with relative density of compaction.

Figure 15–2 shows a direct shear test machine. It consists primarily of a direct shear box, which is split into two halves (top and bottom) and which holds the soil specimen; a proving ring to measure the horizontal load applied to a specimen; two dial gauges (one horizontal and one vertical) to measure the deformation of the soil during the test; and a yoke by which a vertical load can be applied to the soil specimen. A horizontal load to the top half of the shear box is applied by a motor and gear arrangement. In a strain-controlled unit, the rate of movement on the top half of the shear box can be controlled.

Figure 15–3 shows the schematic diagram of the shear box. The shear box is split into two halves—top and bottom. The top and bottom halves of the shear box can be held together by two vertical pins. There is a loading head which can be slipped from the top of the shear box to rest on the soil specimen inside the box. There are also three vertical screws and two horizontal screws on the top half of the shear box.

Procedure

1. Remove the shear box assembly. Back off the three vertical and two horizontal screws. Remove the loading head. Insert the two vertical pins to keep the two halves of the shear box together.
2. Weigh some dry sand in a large porcelain dish, W_1. Fill the shear box with sand in small layers. A tamper may be used to compact the sand layers. The top of the compacted specimen should be about ¼ in. (6.4 mm) below the top of the shear box. Level the surface of the sand specimen.

Figure 15-2. A direct shear test machine.

3. Determine the dimensions of the soil specimen (i.e., length L, width B, and height H of the specimen).
4. Slip the loading head down from the top of the shear box to rest on the soil specimen.
5. Put the shear box assembly in place in the direct shear machine.
6. Apply the desired normal load, N, on the specimen. This can be done by hanging dead weights to the vertical load yoke. The top crossbars will rest on the loading head of the specimen which, in turn, rests on the soil specimen.
   Note: In the equipment shown in Fig. 15–2, the weights of the hanger, the loading head, and the top half of the shear box can be tared. In some other equipment, if taring is not possible, the normal load should be calculated as N = load hanger + weight of yoke + weight of loading head + weight of top half of the shear box.
7. Remove the two vertical pins (which were inserted in Step 1 to keep the two halves of the shear box together).
8. Advance the three vertical screws that are located on the side walls of the top half of the shear box. This is done to separate the two halves of the box. The space between

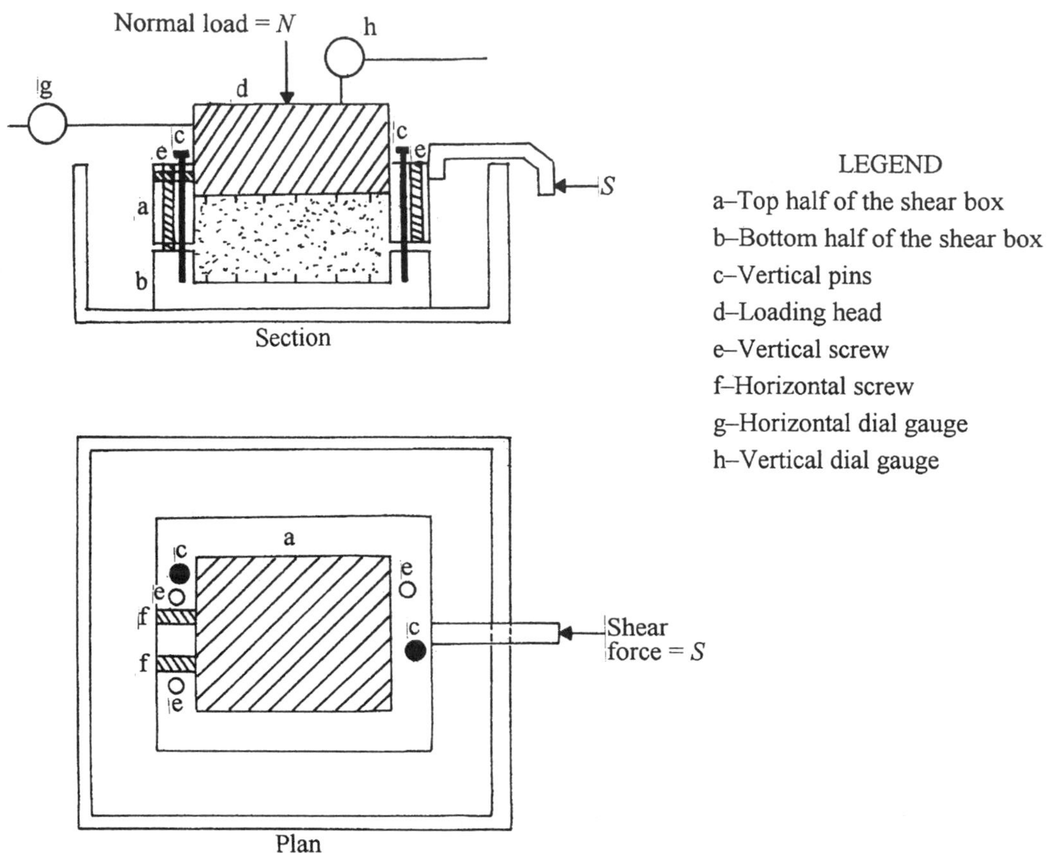

Figure 15–3. Schematic diagram of a direct shear test box.

the two halves of the box should be slightly larger than the largest grain size of the soil specimen (by visual observation).

9. Set the loading head by tightening the two horizontal screws located at the top half of the shear box. Now back off the three vertical screws. After doing this, there will be no connection between the two halves of the shear box except the soil.
10. Attach the horizontal and vertical dial gauges (0.001 in./small div) to the shear box to measure the displacement during the test.
11. Apply horizontal load, S, to the top half of the shear box. The rate of shear displacement should be between 0.1 to 0.02 in./min (2.54 to 0.51 mm/min). For every tenth small division displacement in the horizontal dial gauge, record the readings of the vertical dial gauge and the proving ring gauge (which measures horizontal load, S). Continue this until after
    (a) the proving ring dial gauge reading reaches a maximum and then falls, or
    (b) the proving ring dial gauge reading reaches a maximum and then remains constant.

12. Repeat the test (Steps 1 to 11) at least two more times. For each test, the dry unit weight of compaction of the sand specimen should be the same as that of the first specimen (Step 2).

Calculation

Sample calculations of a direct shear test in sand for one normal load, N, are shown in Tables 15–1 and 15–2. Referring to Table 15–1.

1. Determine the dry unit weight of the specimen, γ_d (Line 6).
   The dry density of the specimen can be calculated as

$$\rho_d = \frac{W_1 - W_2}{LBH} \tag{15.1}$$

If W_1 and W_2 are in grams and L, B and H are in inches,

$$\gamma_d \text{ (lb / ft}^{3)} = \frac{W_1 - W_2}{LBH} \times 3.803 \tag{15.2}$$

2. Calculate the void ratio of the specimen (Line 8)

$$e = \frac{G_s \gamma_w}{\gamma_d} - 1 \tag{15.3}$$

where G_s = specific gravity of soil solids
γ_w = unit weight of water (62.4 lb/ft^3)
and γ_d is in lb/ft^3.

Now refer to Table 15-2

3. Calculate the normal stress, σ', on the specimen (Column 1)

$$\sigma' \text{ (lb / in.}^2) = \frac{\text{Normal load, } N \text{ (lb)}}{(L \text{ in.})(B \text{ in.})} \tag{15.4}$$

4. The horizontal, vertical, and proving ring dial gauge readings are obtained from the test (Columns 2, 3 and 4).
5. For any given set of horizontal and vertical dial gauge readings, calculate the shear force (Column 6)

$$S = \text{(No. of divisions in proving ring dial gauge, i.e., Column 4)} \times \text{(proving ring calibration factor, i.e., Column 5)} \tag{15.5}$$

6. Calculate the shear stress, τ, as (Column 7)

$$\tau \ (\text{lb}/\text{in.}^2) = \frac{\text{shear force, } S}{\text{area of the specimen}} = \frac{S\ (\text{lb})}{(L \text{ in.})(B \text{ in.})} \tag{15.6}$$

*Note:*A separate data sheet has to be used for each test (i.e., for each normal stress, σ').

Graph

1. For each normal stress, plot a graph of τ (Column 7) vs. horizontal displacement (Column 2) (as shown in Fig. 15–4 for the results obtained from Table 15–1). On the bottom of the same graph paper, using the same horizontal scale, plot a graph of verti- cal displacement (Column 3) vs. horizontal displacement (Column 2). There will be at least three such plottings (one for each value of σ'). Determine the shear stresses at failure, s, from each τ vs. horizontal displacement graph (as shown in Fig. 15–4).
2. Plot a graph of shear strength, s, vs. normal stress, σ'. This graph will be a straight line passing through the origin. Figure 15–5 shows such a plot for the sand reported in Tables 15–1 and 15–2. The angle of friction of the soil can be determined from the slope of the straight line plot of s vs. σ' as

$$\phi = \tan^{-1}\left(\frac{s}{\sigma'}\right) \tag{15.7}$$

Table 15–1. Direct Shear Test on Sand
Void Ratio Calculation

Description of soil *Uniform sand* Sample No. *2*

Location *Argonaut Circle*

Tested by ______ Date ______

Item	Quantity
1. Specimen length, L (in.)	*2*
2. Specimen width, B (in.)	*2*
3. Specimen height, H (in.)	*1.31*
4. Mass of porcelain dish + dry sand (before use), W_1 (g)	*540.3*
5. Mass of porcelain dish + dry sand (after use), W_2 (g)	*397.2*
6. Dry unit weight of specimen, $\gamma_d\ (\text{lb}/\text{ft}^3) = \dfrac{W_1 - W_2\ (\text{g})}{LBH\ (\text{in.}^3)} \times 3.808$	*104.0*
7. Specific gravity of soil solids, G_s	*2.66*
8. Void ratio, $e = \dfrac{G_s \gamma_w}{\gamma_d} - 1$ *Note:* $\gamma_w = 62.4$ lb/ft^3; γ_d is in lb/ft^3	*0.596*

Table 15–2. Direct Shear Test on Sand
Stress and Displacement Calculation

Description of soil *Uniform sand* Sample No. *2*

Location *Argonaut Circle*

Normal load, N *56* lb Void ratio, e *0.596*

Tested by ______ Date ______

Normal stress, σ' (lb/in.^{2}) (1)	Horizontal displacement (in.) (2)	Vertical displacement (in.) (3)	No. of div. in proving ring dial gauge (4)	Proving ring calibration factor (lb/div.) (5)	Shear force, S (lb) (6)	Shear stress, τ (lb/in.^{2}) (7)
14	0	0	0	0.31	0	0
14	0.01	+0.001	45	0.31	13.95	3.49
14	0.02	+0.002	76	0.31	23.56	5.89
14	0.03	+0.004	95	0.31	29.76	7.44
14	0.04	+0.006	112	0.31	34.72	8.68
14	0.05	+0.008	124	0.31	38.44	9.61
14	0.06	+0.009	129	0.31	39.99	10.00
14	0.07	+0.010	125	0.31	38.75	9.69
14	0.08	+0.010	119	0.31	36.89	9.22
14	0.09	+0.009	114	0.31	35.34	8.84
14	0.10	+0.008	109	0.31	33.79	8.45
14	0.11	+0.008	108	0.31	33.48	8.37
14	0.12	+0.008	105	0.31	32.55	8.14

* Plus (+) sign means expansion

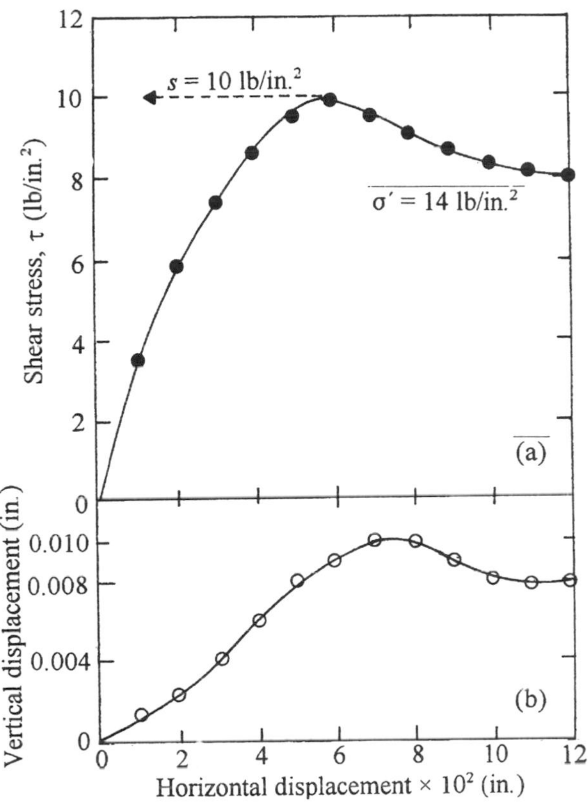

Figure 15–4. Plot of shear stress and vertical displacement vs. horizontal displacement for the direct shear test reported in Tables 15–1 and 15–2.

General Comments

Typical values of the drained angle of friction, ϕ, for sands are given below:

Round-grained sand	ϕ (deg)	Angular-grained sand	ϕ (deg)
Loose	28–32	Loose	30–36
Medium	30–35	Medium	34–40
Dense	34–38	Dense	40–45

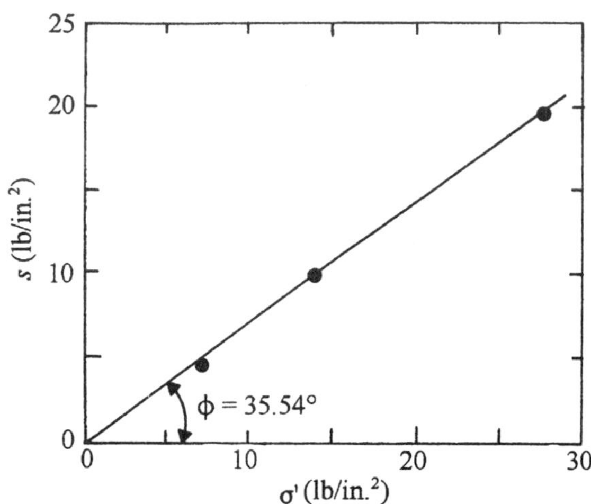

Figure 15–5. Plot of s vs. σ' for the sand reported in Tables 15–1 and 15–2.

Note: The results for tests with $\sigma' = 7$ lb/in.^{2} and 28 lb/in.^{2} are not shown in Table 15–2.

16
Unconfined Compression Test

Introduction

Shear strength of a soil can be given by the Mohr-Coulomb failure criteria as

$$s = c + \sigma + \tan \phi \tag{16.1}$$

where s = shear strength
c = cohesion
σ = normal stress
ϕ = angle of friction.

For *undrained tests of saturated clayey soils* ($\phi = 0$)

$$s = c_u \tag{16.2}$$

where c_u = undrained cohesion (or undrained shear strength).

The unconfined compression test is a quick method of determining the value of c_u for a clayey soil. The unconfined strength is given by the relation [for further discussion see any soil mechanics text, e.g., Das (1994)]

$$q_u = \frac{c_u}{2} \tag{16.3}$$

where q_u = unconfined compression strength.

The unconfined compression strength is determined by applying an axial stress to a cylindrical soil specimen with no confining pressure and observing the axial strains corresponding to various stress levels. The stress at which failure in the soil specimen occurs is referred to as the unconfined compression strength (Figure 16–1). For *saturated* clay specimens, the unconfined compression strength decreases with the increase in moisture content. For *unsaturated* soils, with the dry unit weight remaining constant, the unconfined compression strength decreases with the increase in the degree of saturation.

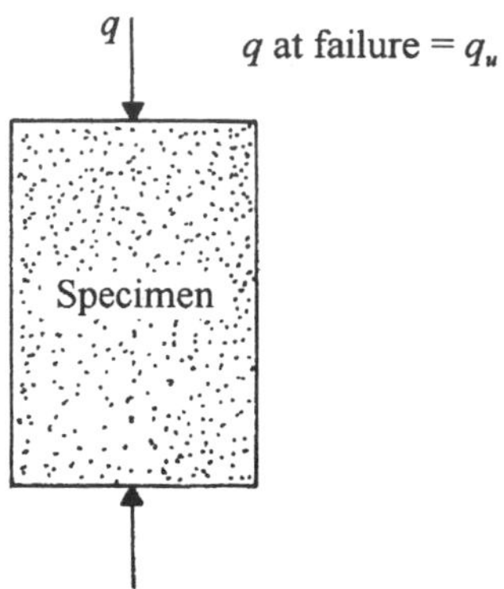

Figure 16–1. Unconfined compression strength—definition

Equipment

1. Unconfined compression testing device
2. Specimen trimmer and accessories (if undisturbed field specimen is used)
3. Harvard miniature compaction device and accessories (if a specimen is to be molded for classroom work)
4. Scale
5. Balance sensitive to 0.01 g
6. Oven
7. Porcelain evaporating dish

Unconfined Compression Test Machine

An unconfined compression test machine in which *strain-controlled* tests can be performed is shown in Fig. 16–2. The machine essentially consists of a top and a bottom loading plate. The bottom of a proving ring is attached to the top loading plate. The top of the proving ring is attached to a cross-bar which, in turn, is fixed to two metal posts. The bottom loading plate can be moved up or down.

Procedure

1. Obtain a soil specimen for the test. If it is an undisturbed specimen, it has to be trimmed to the proper size by using the specimen trimmer. For classroom laboratory work, specimens at various moisture contents can be prepared using a Harvard miniature compaction device.

   The cylindrical soil specimen should have a height-to-diameter (L/D) ratio of between 2 and 3. In many instances, specimens with diameters of 1.4 in. (35.56 mm) and heights of 3.5 in. (88.9 mm) are used.

Figure 16–2. An unconfined compression testing machine.

2. Measure the diameter (D) and length (L) of the specimen and determine the mass of the specimen.
3. Place the specimen centrally between the two loading plates of the unconfined compression testing machine. Move the top loading plate very carefully just to touch the top of the specimen. Set the proving ring dial gauge to zero.

   A dial gauge [each small division of the gauge should be equal to 0.001 in. (0.0254 mm) of vertical travel] should be attached to the unconfined compression testing machine to record the vertical upward movement (i.e., compression of the specimen during testing) of the bottom loading plate. Set this dial gauge to zero.
4. Turn the machine on. Record loads (i.e., proving ring dial gauge readings) and the corresponding specimen deformations. During the load application, the rate of *vertical strain* should be adjusted to ½% to 2% per minute. At the initial stage of the test, readings are usually taken every 0.01 in. (0.254 mm) of specimen deformation. However, this can be varied to every 0.02 in. (0.508 mm) of specimen deformation at a later stage of the test when the load-deformation curve begins to flatten out.
5. Continue taking readings until
   a. Load reaches a peak and then decreases; or

Figure 16-3. A soil specimen after failure

b. Load reaches a maximum value and remains approximately constant thereafter (take about 5 readings after it reaches the peak value); or
c. Deformation of the specimen is past 20% strain before reaching the peak. This may happen in the case of soft clays.

Figure 16–3 shows a soil specimen after failure.

6. Unload the specimen by lowering the bottom loading plate.
7. Remove the specimen from between the two loading plates.
8. Draw a free-hand sketch of the specimen after failure. Show the nature of the failure.
9. Put the specimen in a porcelain evaporating dish and determine the moisture content (after drying it in an oven to a constant weight).

Calculation

For each set of readings (refer to Table 16–1):

1. Calculate the vertical strain (Column 2)

$$\varepsilon = \frac{\Delta L}{L} \tag{16.4}$$

where ΔL = total vertical deformation of the specimen
L = original length of specimen.

2. Calculate the vertical load on the specimen (Column 4).

$$\text{Load} = (\text{proving ring dial reading, i.e. Column 3}) \times (\text{calibration factor}) \tag{16.5}$$

Table 16–1. Unconfined Compression Test

Description of soil *Light brown clay* Specimen No. *3*

Location *Trinity Boulevard*

Moist weight of specimen *149.8* g Moisture content *12* % Length of specimen, L *3* in. Diameter of specimen *1.43* in.

Proving ring calibration factor: 1 div. = *0.264* lb Area, $A_0 = \frac{\pi}{4}D^2$ = *1.605* in.^{2}

Tested by ______ Date ______

Specimen deformation = ΔL (in.) (1)	Vertical strain, $\epsilon = \frac{\Delta L}{L}$ (2)	Proving ring dial reading [No. of small divisions] (3)	Load = Column 3 × calibration factor of proving ring (lb) (4)	Corrected area = $A_c = \frac{A_0}{1-\epsilon}$ (in.^{2}) (5)	Stress, $\sigma = \frac{\text{Column 4}}{\text{Column 5}}$ (lb/in.^{2}) (6)
0	0	0	0	1.605	0
0.01	0.0033	12	3.168	1.611	1.966
0.02	0.0067	38	10.032	1.617	6.205
0.03	0.01	52	13.728	1.622	8.462
0.04	0.013	58	15.312	1.628	9.407
0.06	0.02	67	17.688	1.639	10.793
0.08	0.027	74	19.536	1.650	11.840
0.10	0.033	78	20.592	1.661	12.394
0.12	0.04	81	21.384	1.673	12.782
0.14	0.047	83	21.912	1.685	13.007
0.16	0.053	85	22.440	1.697	13.227
0.18	0.06	86	22.704	1.709	13.288
0.20	0.067	86	22.704	1.721	13.194
0.24	0.08	84	22.176	1.746	12.703
0.28	0.093	83	21.912	1.771	12.370
0.32	0.107	82	21.912	1.798	12.041
0.36	0.12	81	21.384	1.825	11.717

3. Calculate the corrected area of the specimen (Column 5)

$$A_c = \frac{A_0}{1-\varepsilon} \tag{16.6}$$

where A_0 = initial area of cross-section of the specimen $= \frac{\pi}{4} D^2$

4. Calculate the stress, σ, on the specimen (Column 6)

$$\sigma = \frac{\text{Load}}{A_c} = \frac{\text{Column 4}}{\text{Column 5}} \tag{16.7}$$

Graph

Plot the graph of stress, σ (Column 6), vs. axial strain, ϵ, in percent (Column 2 × 100). Determine the peak stress from this graph. This is the unconfined compression strength, q_u, of the specimen. *Note:* If 20% strain occurs before the peak stress, then the stress corresponding to 20% strain should be taken as q_u.

A sample calculation and graph are shown in Table 16–1 and Fig. 16–4.

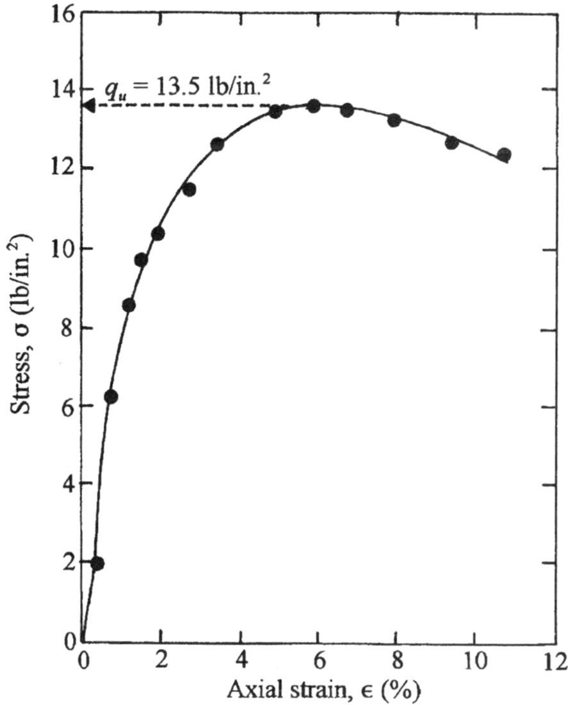

Figure 16–4. Plot of σ vs. ϵ (%) for the test results shown in Table 16–1.

General Comments

1. In the determination of unconfined compression strength, it is better to conduct tests on two to three identical specimens. The average value of q_u is the representative value.
2. Based on the value of q_u, the consistency of a cohesive soil is as follows:

Consistency	q_u (lb/ft^2
Very soft	0–500
Soft	500–1000
Medium	1000–2000
Stiff	2000–4000
Very stiff	4000–8000

3. For many naturally deposited clayey soils, the unconfined compression strength is greatly reduced when the soil is tested after remolding without any change in moisture content. This is referred to as *sensitivity* and can be defined as

$$S_t = \frac{q_{u(\text{undisturbed})}}{q_{u(\text{remolded})}} \tag{16.8}$$

The sensitivity of most clays ranges from 1 to 8. Based on the magnitude of S_t, clays can be described as follows:

Sensitivity, S_t	Description
1–2	Slightly sensitive
2–4	Medium sensitivity
4–8	Very sensitive
8–16	Slightly quick
16–32	Medium quick
32–64	Very quick
> 64	Extra quick

17
Consolidation Test

Introduction

Consolidation is the process of time-dependent settlement of saturated clayey soil when subjected to an increased loading. In this chapter, the procedure of a one-dimensional laboratory consolidation test will be described, and the methods of calculation to obtain the void ratio-pressure curve (e vs. $\log p$), the preconsolidation pressure (p_c), and the coefficient of consolidation (c_v) will be outlined.

Equipment

1. Consolidation test unit
2. Specimen trimming device
3. Wire saw
4. Balance sensitive to 0.01 g
5. Stop watch
6. Moisture can
7. Oven

Consolidation Test Unit

The consolidation test unit consists of a consolidometer and a loading device. The consolidometer can be either (*I*) a floating ring consolidometer (Fig. 17–1*a*) or (*ii*) a fixed ring consolidometer (Fig. 17–1*b*). The floating ring consolidometer usually consists of a brass ring in which the soil specimen is placed. One porous stone is placed at the top of the specimen and another porous tone at the bottom. The soil specimen in the ring with the two porous stones is placed on a base plate. A plastic ring surrounding the specimen fits into a groove on the base plate. Load is applied through a loading head that is placed on the top porous stone. In the floating ring consolidometer, compression of the soil specimen occurs from the

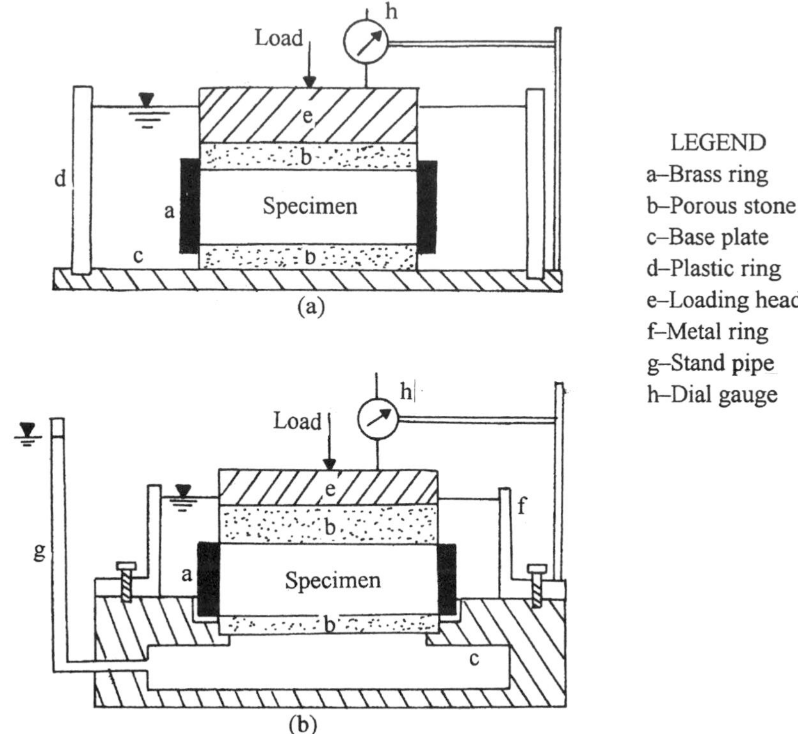

Figure 17–1. Schematic diagram of (a) floating ring consolidometer; (b) fixed ring consolidometer.

top and bottom towards the center. The fixed ring consolidometer essentially consists of the same components, i.e., a hollow base plate, two porous stones, a brass ring to hold the soil specimen, and a metal ring that can be fixed tightly to the top of the base plate. The ring surrounds the soil specimen. A stand pipe is attached to the side of the base plate. This can be used for permeability determination of soil. In the fixed ring consolidometer, the compression of the specimen occurs from the top towards the bottom.

The specifications for the loading devices of the consolidation test unit vary depending upon the manufacturer. Figure 17–2 shows one type of loading device.

During the consolidation test, when load is applied to the soil specimen, the nature of variation of side friction between the surrounding brass ring and the specimen are different for the fixed ring and the floating ring consolidometer, and this is shown in Fig. 17–3. In most cases, a side friction of 10% of the applied load is a reasonable estimate.

Procedure

1. Prepare a soil specimen for the test. The specimen is prepared by trimming an undisturbed natural sample obtained in shelby tubes. The shelby tube sample should be about ¼ in. to ½ in. (6.35 mm to 12.7 mm) larger in diameter than the specimen diameter to be prepared for the test.

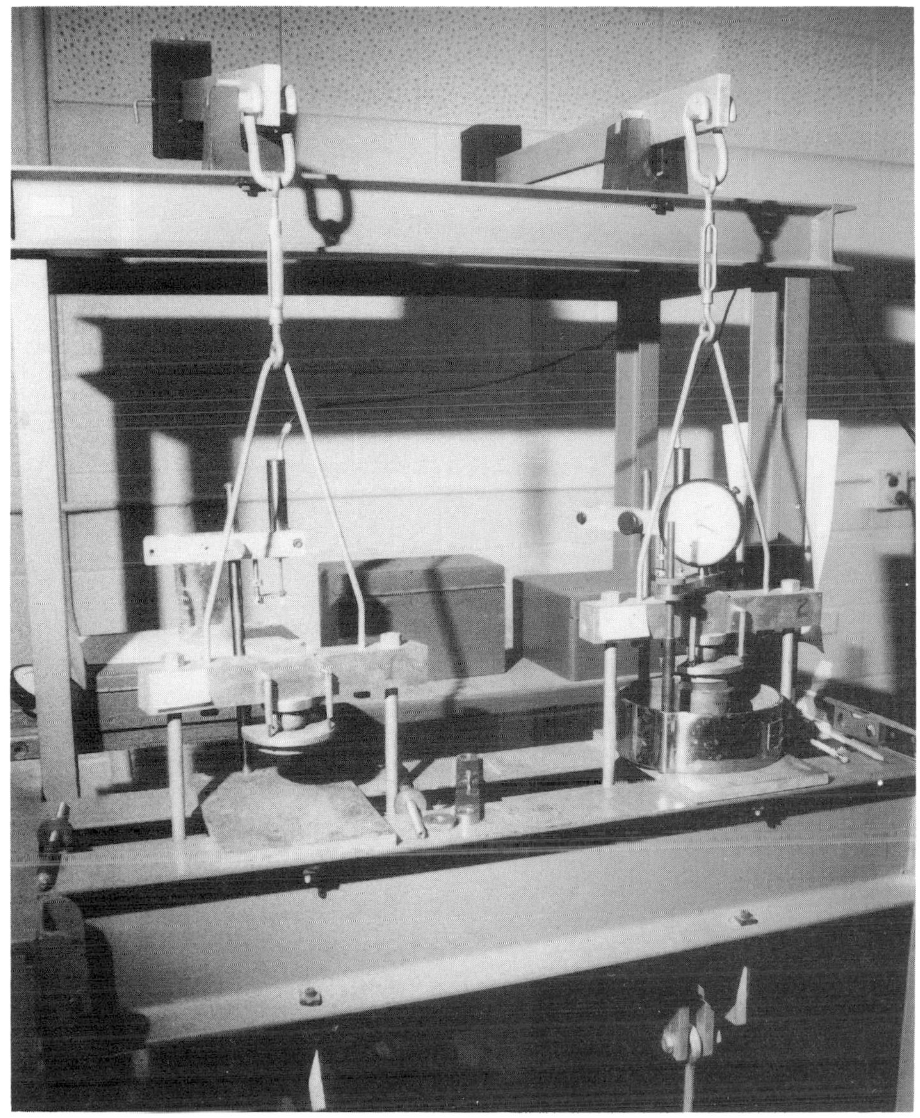

Figure 17–2. Consolidation load assembly. In this assembly, two specimens can be simultaneously tested. Lever arm ratio for loading is 1:10.

Note: For classroom instruction purposes, a specimen can be molded in the laboratory.

2. Collect some excess soil that has been trimmed in a moisture can for moisture content determination.
3. Collect some of the excess soil trimmed in Step 1 for determination of the specific gravity of soil solids, G_s.
4. Determine the mass of the consolidation ring (W_1) in grams.

5. Place the soil specimen in the consolidation ring. Use the wire saw to trim the specimen flush with the top and bottom of the consolidation ring. Record the size of the specimen, i.e., height $[H_{t(i)}]$ and diameter (D).

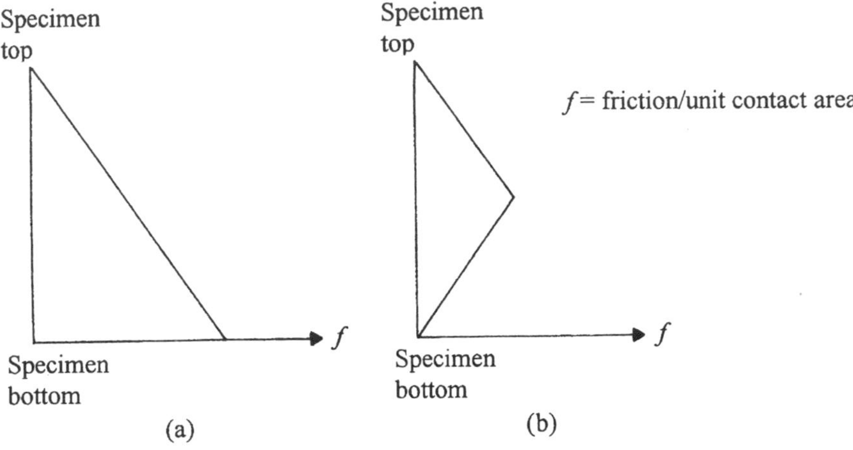

Figure 17–3. Nature of variation of soil-ring friction per unit contact areas in (a) fixed ring consolidometer; (b) floating ring consolidometer.

6. Determine the mass of the consolidation ring and the specimen (W_2) in grams.
7. Saturated the lower porous stone on the base of the consolidometer.
8. Place the soil specimen in the ring over the lower porous stone.
9. Place the upper porous stone on the specimen in the ring.
10. Attach the top ring to the base of the consolidometer.
11. Add water to the consolidometer to submerge the soil and keep it saturated. In the case of the fixed ring consolidometer, the outside ring (which is attached to the top of the base) and the stand pipe connection attached to the base should be kept full with water. This needs to be done for the *entire period of the test*.
12. Place the consolidometer in the loading device.
13. Attach the vertical deflection dial gauge to measure the compression of soil. It should be fixed in such as way that the dial is at the beginning of its release run. The dial gauge should be calibrated to read as 1 small division = 0.0001. (0.00254 mm).
14. Apply load to the specimen such that the magnitude of pressure, p, on the specimen is ½ ton/ft^2 (45.88 kN/m^2). Take the vertical deflection dial gauge readings at the following times, t, counted from the time of load application—0 min., 0.25 min. 1 min., 2.25 min., 4 min., 6.25 min., 9 min., 12.25 min., 20.25 min., 25 min., 36 min., 60 min., 120 min., 240 min., 480 min., and 1440 min. (24 hr.).
15. The next day, add more load to the specimen such that the total magnitude of pressure on the specimen becomes 1 ton/ft^2 (95.76 kN/m^2). Take the vertical dial gauge reading at similar time intervals as stated in Step 14. *Note:* Here we have $\Delta p / p = 1$ (where Δp = increase in pressure and p = the pressure before the increase).

16. Repeat Step 15 for soil pressure magnitudes of 2 ton/ft^2 (299.52 kN/m^2), 4 ton/ft^2 (383.04 kN/m^2) and 8 ton/ft^2 (766.08 kN/m^2). *Note:* $\Delta p/p = 1$.
17. At the end of the test, remove the soil specimen and determine its moisture content.

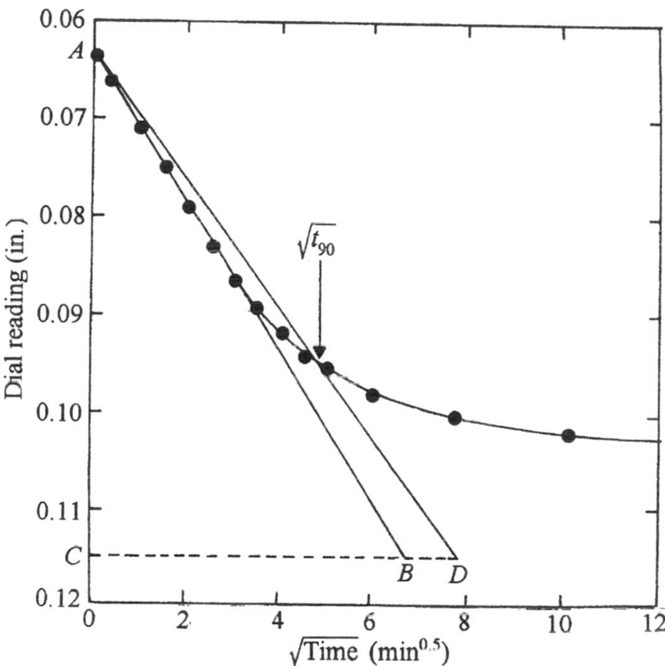

Figure 17–4. Plot of dial reading vs. $\sqrt{\text{time}}$ for the test results given in Table 17–1. Determination of t_{90} by square-root-of-time method.

Calculation and Graph

The calculation procedure for the test can be explained with reference to Tables 17–1 and 17–2 and Figs. 17–4, 17–5 and 17–6, which show the laboratory test results for a light brown clay.

1. Collect all of the time vs. vertical dial readings data. Table 17–1 shows the results of a pressure increase from $p = 2$ ton/ft^2 to $p + \Delta p = 4$ ton/ft^2.
2. Determine the time for 90% primary consolidation, t_{90}, from each set of time vs. vertical dial readings. An example of this is shown in Fig. 17–4, which is a plot of the results of vertical dial reading vs. $\sqrt{\text{time}}$ given in Table 17–1. Draw a tangent *AB* to the initial consolidation curve. Measure the length *BC*. The abscissa of the point of intersection of the line *AD* with the consolidation curve will give $\sqrt{t_{90}}$. In Fig. 17–4, $\sqrt{t_{90}} = 4.75$ min.$^{0.5}$, so $t_{90} = (4.75)^2 = 22.56$ min. This technique is referred to as the *square-root-of-time fitting method* (Taylor, 1942).
3. Determine the time for 50% primary consolidation, t_{50}, from each set of time vs. vertical dial readings. The procedure for this is shown in Fig. 17–5, which is a semilog plot (vertical dial reading in natural scale and time in log scale) for the set of readings shown in Table 17–1. Project the straight line portion of the primary consolidation

Table 17–1. Consolidation Test
Time vs. Vertical Dial Reading

Description of soil *Light brown clay*

Location *Summit Drive*

Tested by ______________________ Date ______________

Pressure on specimen *4* lb/ft^2			Pressure on specimen _____ lb/ft^2		
Time after load application, t (min.)	$\sqrt{t}$ (min.)$^{0.5}$	Vertical dial reading (in.)	Time after load application, t (min.)	$\sqrt{t}$ (min.)$^{0.5}$	Vertical dial reading (in.)
0	*0*	*0.0638*			
0.25	*0.5*	*0.0654*			
1.0	*1.0*	*0.0691*			
2.25	*1.5*	*0.0739*			
4.0	*2.0*	*0.0795*			
6.25	*2.5*	*0.0833*			
9.0	*3.0*	*0.0868*			
12.25	*3.5*	*0.0898*			
16.0	*4.0*	*0.0922*			
20.25	*4.5*	*0.0941*			
25	*5.0*	*0.0954*			
36	*6.0*	*0.0979*			
60	*7.75*	*0.1004*			
120	*10.95*	*0.1019*			
240	*15.49*	*0.1029*			
480	*21.91*	*0.1048*			
1440	*37.95*	*0.1059*			

Table 17–1. Consolidation Test
Void Ratio-Pressure and Coefficient of Consolidation Calculation

Description of soil *Light brown clay* Location ______

Specimen diameter *2.5* in. Initial specimen height, $H_{t(i)}$ *1* in. Height of solids, H_s *1.356* cm = *0.539* in.

Moisture Content: Beginning of test *30.8* % End of test *32.1* % Weight of dry soil specimen *116.74* g G_s *2.72*

Tested by ______ Date ______

Pressure, p (ton/ft²) (1)	Final dial reading (in.) (2)	Change in specimen height ΔH (in.) (3)	Final specimen height, $H_{t(f)}$ (in.) (4)	Height of void, H_v (in.) (5)	Final void ratio, e (6)	Average height during consolidation, $H_{t(av)}$ (in.) (7)	Fitting time (sec)		c_v from × 10^3 (in.²/sec)	
							t_{90} (8)	t_{50} (9)	t_{90} (10)	t_{50} (11)
0	0.200		1.000	0.4610	0.855					
		0.0083				0.9959	302	68.7	0.696	0.711
½	0.0283		0.9917	0.4527	0.840					
		0.0073				0.9881	308	56.0	0.672	0.859
1	0.0356		0.9844	0.4454	0.826					
		0.0282				0.9703	492	144	0.406	0.322
2	0.0638		0.9562	0.4172	0.774					
		0.0421				0.9352	1102	294	0.168	0.147
4	0.1059		0.9141	0.3751	0.696					
		0.0455				0.8914	1354	240	0.124	0.163
8	0.1514		0.8686	0.3296	0.612					

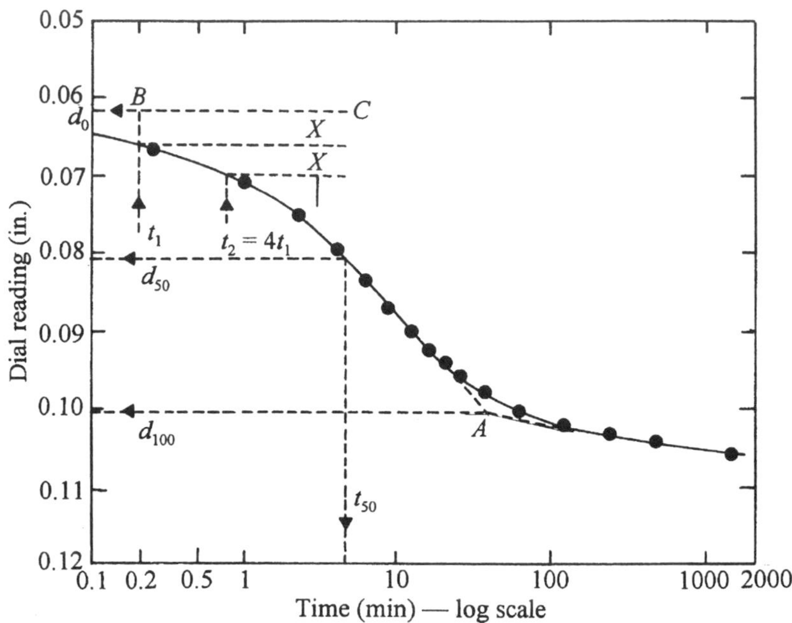

Figure 17–5. Logarithm of time curve fitting method for the laboratory results given in Table 17–1.

downward and the straight line portion of the secondary consolidation backward. The point of intersection of these two lines is A. The vertical dial reading corresponding to A is d_{100} (dial reading at 100% primary consolidation). Select times t_1 and $t_2 = 4t_1$. (*Note:* t_1 and t_2 should be within the top curved portion of the consolidation plot.) Determine the difference in dial readings, X, between times t_1 and t_2. Plot line BC, which is vertically X distance above the point on the consolidation curve corresponding to time t_1. The vertical dial gauge corresponding to line BC is d_0, i.e., the reading for 0% consolidation. Determine the dial gauge reading corresponding to 50% primary consolidation as

$$d_{50} = \frac{d_0 + d_{100}}{2} \tag{17.1}$$

The time corresponding to d_{50} on the consolidation curve is t_{50}. This is the logarithm-of-time curve fitting method (Casagrande and Fadum, 1940). In Figure 17–5, t_{50} = 14.9 min.

4. Complete the experimental data in Columns 1, 2, 8 and 9 of Table 17–2. Columns 1 and 2 are obtained from time-dial reading tables (such as Table 17–1), and Columns 8 and 9 are obtained from Steps 2 and 3, respectively.
5. Determine the height of solids (H_s) of the specimen in the mold as (see top of Table 17–2)

$$H_s = \frac{W_s}{\left(\frac{\pi}{4}D^2\right)G_s\rho_w} \tag{17.2}$$

where W_s = dry mass of soil specimen
D = diameter of the specimen
G_s = specific gravity of soil solids
ρ_w = density of water.

6. In Table 17–2, determine the change in height, ΔH, of the specimen due to load increments from p to $p + \Delta p$ (Column 3). For example,
p = ½ ton/ft^2, final dial reading = 0.0283 in.
$p + \Delta p$ = 1 ton/ft^2, final dial reading = 0.0356 in.
Thus
ΔH = 0.0356 − 0.0283 = 0.0073 in.
7. Determine the final specimen height, $H_{t(f)}$, at the end of consolidation due to a given load (Column 4 in Table 17–2). For example, in Table 17–2 $H_{t(f)}$ at p = ½ ton/ft^2 is 0.9917. ΔH from p = ½ ton/ft^2 and 1 ton/ft^2 is 0.0073. So $H_{t(f)}$ at p = 1 ton/ft^2 is 0.9917 − 0.0073 = 0.9844 in.
8. Determine the height of voids, H_v, in the specimen at the end of consolidation due to a given loading, p, as (see Column 5 in Table 17–2)

$$H_v = H_{t(f)} - H_s \tag{17.3}$$

9. Determine the final void ratio at the end of consolidation for each loading, p, as (see Column 6, Table 17–2)

$$e = \frac{H_v}{H_s} = \frac{\text{Column 5}}{H_s} \tag{17.4}$$

10. Determine the average specimen height, $H_{t(\text{av})}$, during consolidation for each incremental loading (Column 7, Table 17–2). For example, in Table 17–2, the value of $H_{t(\text{av})}$ between p = ½ ton/ft^2 and p = 1 ton/ft^2 is

$$\frac{H_{t(f)}\ at\ p = \frac{1}{2}\ ton/ft^2 + H_{t(f)}\ at\ p = 1\ ton/ft^2}{2} = \frac{0.9917 + 0.9844}{2} = 0.9881\ in.$$

11. Calculate the coefficient of consolidation, c_v (Column 10, Table 17–2), from t_{90} (Column 8) as

$$T_v = \frac{c_v t}{H^2} \tag{17.5}$$

where T_v = time factor t_{90} = 0.848
H = maximum length of drainage path = $\dfrac{H_{t(\text{av})}}{2}$ (since the specimen is drained at top and bottom)

Thus

$$c_v = \frac{0.848 H_{t(\text{av})}^2}{4t_{90}} \tag{17.6}$$

12. Calculate the coefficient of consolidation, c_v (Column 11, Table 17–2), from t_{50} (Column 9) as

$$T_{v(50\%)} = 0.197 = \frac{c_v t_{50}}{H^2} = \frac{c_v t_{50}}{\left[\dfrac{H_{t(\text{av})}}{2}\right]^2}$$

$$c_v = \frac{0.197 H_{t(\text{av})}^2}{4t_{50}} \tag{17.7}$$

For example, from p = ½ ton/ft^2 to p = 1 ton/ft^2, $H_{t(\text{av})}$ = 0.9881 in.; t_{50} = 56.0 s. So

$$c_v = \frac{0.197(0.9881)^2}{4(56)} = 0.859 \times 10^{-3} \text{ in.}^2/\text{s}$$

13. Plot a semilogarithmic graph of pressure vs. final void ratio (Column 1 vs. Column 6, Table 17–2). Pressure, p, is plotted on the log scale and the final void ratio on the linear scale. As an example, the results of Table 17–2 are plotted in Fig. 17–6.
Note: The plot has a curved upper portion and, after that, e vs. log p has a linear relationship.
14. Calculate the compression index, C_c. This is the slope of the linear portion of the e vs. log p plot (Step 13). In Fig. 17–6

$$C_c = \frac{e_1 - e_2}{\log \dfrac{p_2}{p_1}} = \frac{0.696 - 0.612}{\log \dfrac{8}{4}} = 0.279$$

15. On the semilogarithmic graph (Step 13), using the same horizontal scale (the scale for p), plot the values of c_v (Column 10 and 11, Table 17–2). As an example, the

values determined in Table 17–2 are plotted in Fig.17–6.

Note: c_v is plotted on the linear scale corresponding to the average value of p, i.e.,

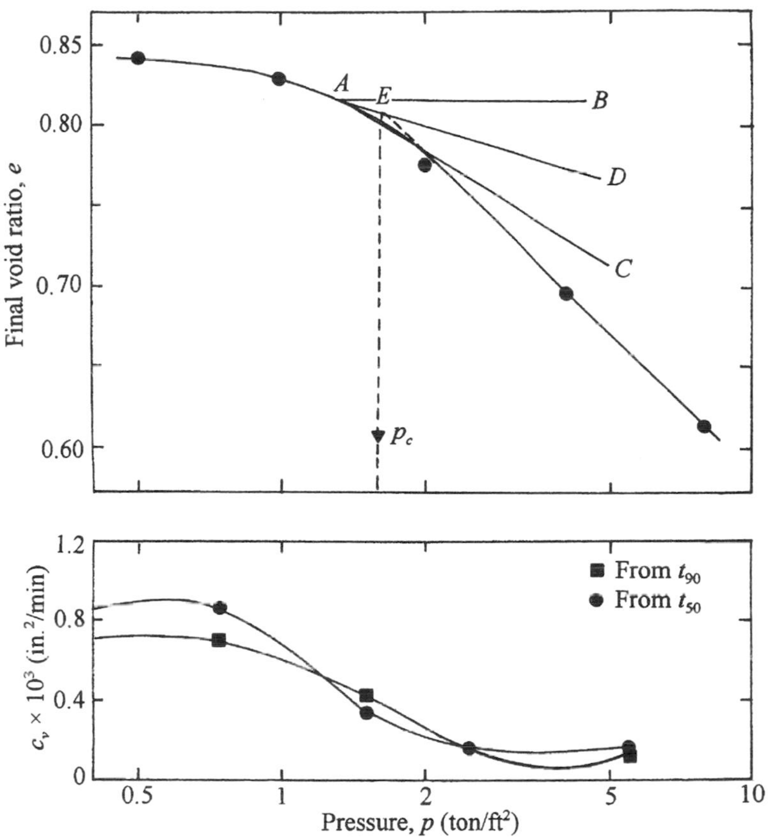

Figure 17–6. Plot of void ratio and the coefficient of consolidation against pressure for the soil reported in Table 17–2.

$$\frac{p_1 + p_2}{2}$$

16. Determine the *preconsolidation pressure, p_c*. The procedure can be explained with the aid of the *e*-log *p* graph shown in Fig. 17–6 (Casagrande, 1936). First, determine point *A*, which is the point on the *e*-log *p* plot that has the smallest radius of curvature. Draw a horizontal line *AB*. Draw a line *AD* which is the *bisector* of angle *BAC*. Project the straight line portion of the *e*-log *p* plot backwards to meet line *AD* at *E*. The pressure corresponding to point *E* is the preconsolidation pressure. In Fig. 17–6, p_c = 1.6 ton/ft^2.

General Comments

The magnitude of the compression index, C_c, varies from soil to soil. Many correlations for C_c have been proposed in the past for various types of soils. A summary of these correlations is given by Rendon-Herrero (1980). Following is a list of some of these correlations.

Correlation	Region of applicability
$C_c = 0.007(LL - 7)$	Remolded clay
$C_c = 0.009(LL - 10)$	Undisturbed clays
$C_c = 1.15(e_0 - 0.27)$	All clays
$C_c = 0.0046(LL - 9)$	Brazilian clays
$C_c = 0.208e_0 + 0.0083$	Chicago clays

Note: LL = liquid limit
e_0 = *in situ* void ratio

18

Triaxial Tests in Clay

Introduction

In Chapters 15 and 16, some aspects of the shear strength for soil were discussed. The triaxial compression test is a more sophisticated test procedure for determining the shear strength of soil. In general, with triaxial equipment, three types of common tests can be conducted, and they are listed below. Both the unconsolidated-undrained test and the consolidated-undrained test will be described in this chapter.

Type of Test	Parameters Determined
Unconsolidated-undrained (*U-U*)	c_u ($\phi = 0$)
Consolidated-drained (*C-D*)	c, ϕ
Consolidated-undrained (*C-U*)	c, ϕ, $\overline{A}$

c_u = undrained cohesion
c = cohesion
ϕ = drained angle of friction
$\overline{A}$ = pore water pressure parameter
Note: $s = c + \sigma'$ tab ϕ (c = cohesion, σ' = effective normal stress).
For undrained condition, $\phi = 0$; $s - c_u$ [Eq. (16.2)]

Equipment

1. Triaxial cell
2. Strain-controlled compression machine
3. Specimen trimmer
4. Wire saw
5. Vacuum source
6. Oven

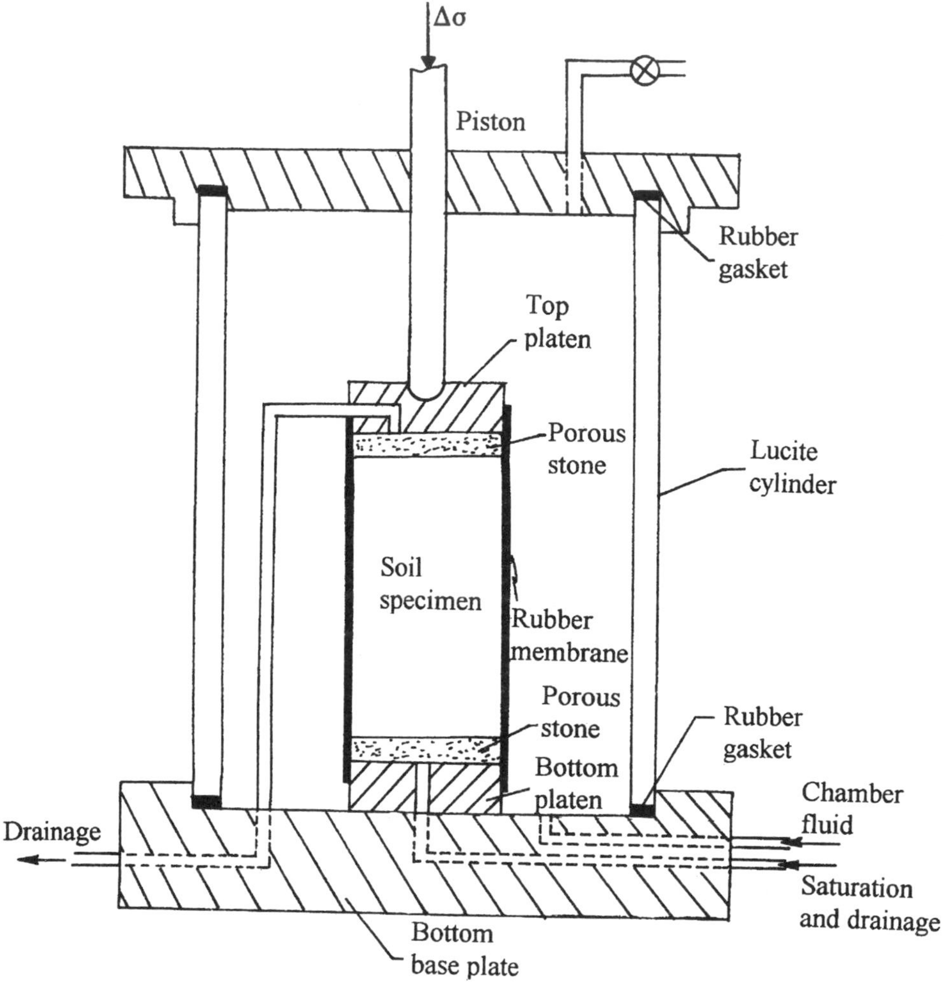

Figure 18–1. Schematic diagram of triaxial cell.

7. Calipers
8. Evaporating dish
9. Rubber membrane
10. Membrane stretcher

Triaxial Cell and Loading Arrangement

Figure 18–1 shows the schematic diagram of a triaxial cell. It consists mainly of a bottom base plate, a Lucite cylinder and a top cover plate. A bottom platen is attached to the base plate. A porous stone is placed over the bottom platen, over which the soil specimen is placed. A porous stone and a platen are placed on top of the specimen. The specimen is enclosed inside a thin rubber membrane. Inlet and outlet tubes for specimen saturation and drainage are provided through the base plate. Appropriate valves to these tubes are attached to shut off the openings when desired. There is an opening in the base plate through which water (or glycerine) can be allowed to flow to fill the cylindrical chamber. A hydrostatic chamber pressure, σ_3, can be applied to the specimen through the chamber fluid. An added axial stress, $\Delta\sigma$, applied to the top of the specimen can be provided using a piston.

During the test, the triaxial cell is placed on the platform of a strain-controlled compression machine. The top of the piston of the triaxial chamber is attached to a proving ring. The proving ring is attached to a crossbar that is fixed to two metal posts. The platform of the compression machine can be raised (or lowered) at desired rates, thereby raising (or lowering) the triaxial cell. During compression, the load on the specimen can be obtained from the proving ring readings and the corresponding specimen deformation from a dial gauge [1 small div. = 0.001 in. (0.0254 mm)].

The connections to the soil specimen can be attached to a burette or a pore-water pressure measuring device to measure, respectively, the volume change of the specimen or the excess pore water pressure during the test.

Triaxial equipment is costly, depending on the accessories attached to it. For that reason, general procedures for tests will be outlined here. For detailed location of various components of the assembly, students will need the help of their instructor.

Triaxial Specimen

Triaxial specimens most commonly used are about 2.8 in. in diameter × 6.5 in. in length (71.1 mm diameter × 165.1 mm length) or 1.4 in. in diameter × 3.5 in. in length (35.6 mm diameter × 88.9 length). In any case, the length-to-diameter ratio (L/D) should be between 2 and 3. For tests on undisturbed natural soil samples collected in shelby tubes, a specimen trimmer may need to be used to prepare a specimen of desired dimensions. Depending on the triaxial cell at hand, for classroom use, remolded specimens can be prepared with Harvard miniature compaction equipment.

After the specimen is prepared, obtain its length (L_0) and diameter (D_0). The length should be measured four times about 90 degrees apart. The average of these four values should be equal to L_0. To obtain the diameter, take four measurements at the top, four at the middle and four at the bottom of the specimen. The average of these twelve measurements is D_0.

Placement of Specimen in the Triaxial Cell

1. Boil the two porous stones to be used with the specimen.
2. De-air the lines connecting the base of the triaxial cell.
3. Attach the bottom platen to the base of the cell.
4. Place the bottom porous stone (moist) over the bottom platen.
5. Take a thin rubber membrane of appropriate size to fit the specimen snugly. Take a membrane stretcher, which is a brass tube with an inside diameter of about ¼ in. (≈ 6 mm) larger than the specimen diameter (Figure 18–2). The membrane stretcher can be connected to a vacuum source. Fit the membrane to the inside of the membrane stretcher and lap the ends of the membrane over the stretcher. Then apply the vacuum. This will make the membrane form a smooth cover inside the stretcher.

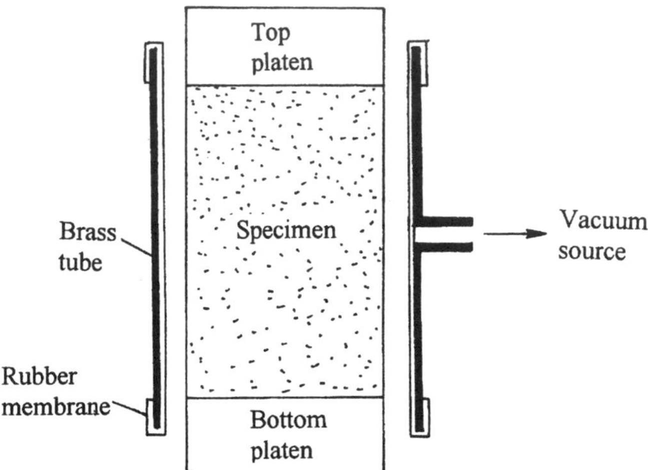

Figure 18–2. Membrane stretcher.

6. Slip the soil specimen to the inside of the stretcher with the membrane (Step 5). The inside of the membrane may be moistened for ease in slipping the specimen in. Now release the vacuum and unroll the membrane from the ends of the stretcher.
7. Place the specimen (Step 6) on the bottom porous stone (which is placed on the bottom platen of the triaxial cell) and stretch the bottom end of the membrane around the porous stone and bottom platen. At this time, place the top porous stone (moist) and the top platen on the specimen, and stretch the top of the membrane over it. For air-tight seals, it is always a good idea to apply some silicone grease around the top and bottom platens before the membrane is stretched over them.
8. Using some rubber bands, tightly fasten the membrane around the top and bottom platens.
9. Connect the drainage line leading from the top platen to the base of the triaxial cell.
10. Place the Lucite cylinder and the top of the triaxial cell on the base plate to complete the assembly.

    Note:

    1. In the triaxial cell, the specimen can be saturated by connecting the drainage line leading to the bottom of the specimen to a saturation reservoir. During this process, the drainage line leading from the top of the specimen is kept open to the atmosphere. The saturation of clay specimens takes a fairly long time.
    2. For the unconsolidated-undrained test, if the specimen saturation is not required, nonporous plates can be used instead of porous stones at the top and bottom of the specimen.

Unconsolidated-Undrained Test

Procedure

1. Place the triaxial cell (with the specimen inside it) on the platform of the compression machine.
2. Make proper adjustments so that the piston of the triaxial cell just rests on the top platen of the specimen
3. Fill the chamber of the triaxial cell with water. Apply a hydrostatic pressure, σ_3, to the specimen through the chamber fluid. *Note:* All drainage to and from the specimen should be closed *now* so that drainage from the specimen does not occur.
4. Check for proper contact between the piston and the top platen on the specimen. Zero the dial gauge of the proving ring and the gauge used for measurement of the vertical compression of the specimen. Set the compression machine for a strain rate of about 0.5% per minute, and turn the switch on.
5. Take initial proving ring dial readings for vertical compression intervals of 0.01 in. (0.254 mm). This interval can be increase to 0.02 in. (0.508 mm) or more later when the rate of increase of load on the specimen decreases. The proving ring readings will increase to a peak value and then may decrease or remain approximately constant. Take about four to five readings after the peak point.
6. After completion of the test, reverse the compression machine, lower the triaxial cell, and then turn off the machine. Release the chamber pressure and drain the water in the triaxial cell. Then remove the specimen and determine its moisture content.

Calculation

The calculation procedure can be explained by referring to Tables 18–1 and 18–2, which present the results of an unconsolidated-undrained triaxial test on a dark brown silty clay specimen. Referring to Table 18–1

1. Calculate the final moisture content of the specimen, w, as (Line 3)

$$w\,(\%) = \frac{\text{moist mass of specimen, } W_1 - \text{dry mass of specimen, } W_2}{\text{dry mass of specimen, } W_2}(100)$$

$$= \frac{\text{Line 1} - \text{Line 2}}{\text{Line 2}}(100) \qquad (18.1)$$

2. Calculate the initial area of the specimen as (Line 6)

$$A_0 = \frac{\pi}{4} D_0^2 = \frac{\pi}{4}(\text{Line 5})^2 \qquad (18.2)$$

3. Now, refer to Table 18–2, calculate the vertical strain as (Column 2)

$$\varepsilon = \frac{\Delta L}{L_0} = \frac{\text{Column 1}}{\text{Line 4, Table 18-1}} \tag{18.3}$$

where ΔL = total deformation of the specimen at any time.

Table 18–1. Unconsolidated-Undrained Triaxial Test Preliminary Data

Description of soil *Dark brown silty clay* Specimen No. *8*

Location ______

Tested by ______ Date ______

Item	Quantity
1. Moist mass of specimen (end of test), W_1	*185.68 g*
2. Dry mass of specimen, W_2	*151.80 g*
3. Moisture content (end of test), $w\ (\%) = \frac{W_1 - W_2}{W_2} \times 100$	*22.3%*
4. Initial average length of specimen, L_0	*3.52 in.*
5. Initial average diameter of specimen, D_0	*1.41 in.*
6. Initial area, $A_0 = \frac{\pi}{4} D^2$	*1.56 in.²*
7. Specific gravity of soil solids, G_s	*2.73*
8. Final degree of saturation	*98.2%*
9. Cell confining pressure, σ_3	*15 lb/in.²*
10. Proving ring calibration factor	*0.37 lb/div.*

Table 18–2. Unconsolidated-Undrained Triaxial Test Axial Stress-Strain Calculation

Specimen deformation $= \Delta L$ (in.) (1)	Vertical strain, $\epsilon = \frac{\Delta L}{L_0}$ (2)	Proving ring dial reading (Number of small divisions) (3)	Piston load, P (Column 3 × calibration factor) lb (4)	Corrected area, $A = \frac{A_0}{1-\epsilon}$ (in.^{2}) (5)	Deviatory stress, $\Delta\sigma = \frac{P}{A}$ (lb/in.^{2}) (6)
0	0	0	0	1.560	0
0.01	0.0028	3.5	1.295	1.564	0.828
0.02	0.0057	7.5	2.775	1.569	1.769
0.03	0.0085	11	4.07	1.573	2.587
0.04	0.0114	14	5.18	1.578	3.28
0.05	0.0142	18	6.66	1.582	4.210
0.06	0.0171	21	7.77	1.587	4.896
0.10	0.0284	31	11.47	1.606	7.142
0.14	0.0398	38	14.06	1.625	8.652
0.18	0.0511	44	16.28	1.644	9.903
0.22	0.0625	48	17.76	1.664	10.673
0.26	0.0739	52	19.24	1.684	11.425
0.30	0.0852	53	19.61	1.705	11.501
0.35	0.0994	52	19.24	1.735	11.109
0.40	0.1136	50	18.5	1.760	10.511
0.45	0.1278	49	18.13	1.789	10.134
0.50	0.1420	49	18.13	1.818	9.970

4. Calculate the piston load on the specimen (Column 4) as

$$P = (\text{proving ring dial reading}) \times (\text{calibration factor}) = (\text{Column 3}) \times (\text{Line 10, Table 18-1}) \quad (18.4)$$

5. Calculate the corrected area, A, of the specimen as (Column 5)

$$A = \frac{A_0}{1-\varepsilon} = \frac{\text{Line 6, Table 18}-1}{1 - \text{Column 2}} \quad (18.5)$$

6. Calculate the deviatory stress (or piston stress), $\Delta\sigma$, as (Column 6)

$$\Delta\sigma = \frac{P}{A} = \frac{\text{Column 4}}{\text{Column 5}} \quad (18.6)$$

Graph

1. Draw a graph of the axial strain (%) vs. deviatory stress ($\Delta\sigma$). As an example, the results of Table 18–2 are plotted in Fig. 18–3. From this graph, obtain the value of $\Delta\sigma$ at failure ($\Delta\sigma = \Delta\sigma_f$).
2. The minor principal stress (*total*) on the specimen at failure is σ_3 (i.e., the chamber confining pressure). Calculate the major principal stress (*total*) at failure as

$$\sigma_1 = \sigma_3 + \Delta\sigma_f$$

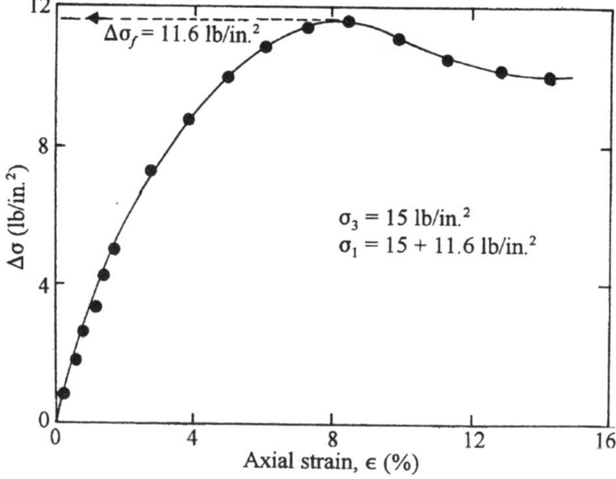

Figure 18–3. Plot of $\Delta\sigma$ against axial strain for the test reported in Table 18–2.

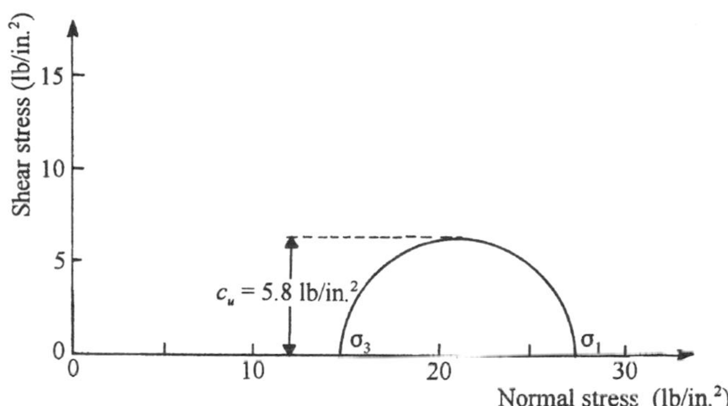

Figure 18–4. Total stress Mohr's circle at failure for test of Table 18–2 and Fig. 18–3.

3. Draw a Mohr's circle with σ_1 and σ_3 as the major and minor principal stresses. The radius of the Mohr's circle is equal to c_u. The results of the test reported in Table 18–2 and Fig. 18–3 are plotted in Fig.18–4.

General Comments

1. For saturated clayey soils, the unconfined compression test (Chapter 16) is a special case of the *U-U* test discussed previously. For the unconfined compression test, $\sigma_3 = 0$. However, the quality of results obtained from *U-U* tests is superior.
2. Figure 18–5 shows the nature of the Mohr's envelope obtained from *U-U* tests with varying degrees of saturation. For *saturated specimens*, the value of $\Delta\sigma_f$, and thus c_u, is constant irrespective of the chamber confining pressure, σ_3. So the Mohr's envelope is a horizontal line ($\phi = 0$). For soil specimens with degrees of saturation less than 100%, the Mohr's envelope is curved and falls above the $\phi = 0$ line.

Consolidated-Undrained Test

Procedure

1. Place the triaxial cell with the saturated specimen on the compression machine platform and make adjustments so that the piston of the cell makes contact with the top platen of the specimen.
2. Fill the chamber of the triaxial cell with water, and apply the hydrostatic pressure, σ_3, to the specimen through the fluid.

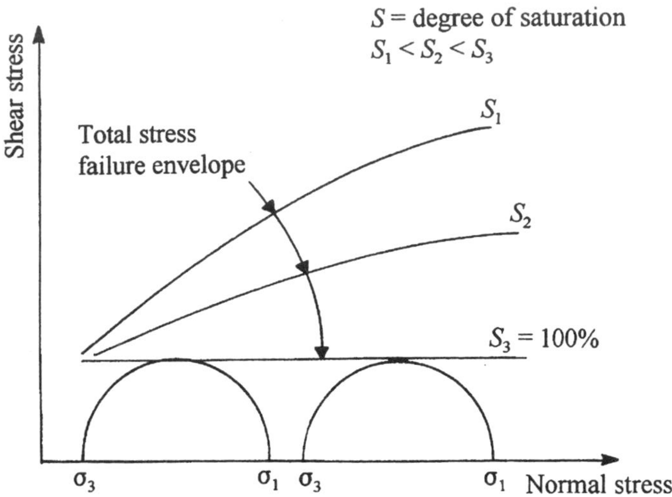

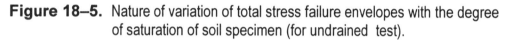

Figure 18–5. Nature of variation of total stress failure envelopes with the degree of saturation of soil specimen (for undrained test).

3. The application of the chamber pressure, σ_3, will cause an increase in the pore water pressure in the specimen. For consolidation connect the drainage lines from the specimen to a calibrated burette and leave the lines open. When the water level in the burette becomes constant, it will indicate that the consolidation is complete. For a saturated specimen, the volume change due to consolidation is equal to the volume of water drained from the burette. Record the volume of the drainage (ΔV).
4. Now connect the drainage lines to the pore-pressure measuring device.
5. Check the contact between the piston and the top platen. Zero the proving ring dial gauge and the dial gauge, which measures the axial deformation of the specimen.
6. Set the compression machine for a strain rate of about 0.5% per minute, and turn the switch on. When the axial load on the specimen is increased, the pore water pressure in the specimen will also increase. Record the proving ring dial gauge reading and the corresponding excess pore water pressure (Δu) in the specimen for every 0.01 in. (0.254 mm) or less of axial deformation. The proving ring dial gauge reading will increase to a maximum and then decrease or remain approximately constant. Take at least four to five readings after the proving ring dial gauge reaches the maximum value.
7. At the completion of the test, reverse the compression machine and lower the triaxial cell. Shut off the machine. Release the chamber pressure, σ_3, and drain the water out of the triaxial cell.
8. Remove the tested specimen from the cell and determine its moisture content.
9. Repeat the test on one or two more similar specimens. Each specimen should be tested at a different value of σ_3.

Calculation and Graph

The procedure for making the required calculations and plotting graphs can be explained by referring to Tables 18–3 and 18–4 and Figs. 18–6 and 18–7. First, referring to Table 18–3,

1. Calculate the initial area of the specimen as (Line 5)

$$A_0 = \frac{\pi}{4} D_0^2 = \frac{\pi}{4} (\text{Line } 4)^2 \tag{18.8}$$

2. Calculate the initial volume of the specimen as (Line 6)

$$V_0 = A_0 L_0 = (\text{Line } 5) \times (\text{Line } 3) \tag{18.9}$$

3. Calculate the volume of the specimen after consolidation as (Line 9)

$$V_c = V_0 - \Delta V = (\text{Line } 6) - (\text{Line } 8) \tag{18.10}$$

where V_c = final volume of the specimen.

4. Calculate the length, L_c (Line 10), and cross-sectional area, A_c (Line 11) of the specimen after consolidation as

$$L_c = L_0 \left(\frac{V_c}{V_0} \right)^{1/3} = (\text{Line } 3) \left(\frac{\text{Line } 9}{\text{Line } 6} \right)^{1/3} \tag{18.11}$$

and

$$A_c = A_0 \left(\frac{V_c}{V_0} \right)^{2/3} = (\text{Line } 5) \left(\frac{\text{Line } 9}{\text{Line } 6} \right)^{2/3} \tag{18.12}$$

Now refer to Table 18–4.

5. Calculate the axial strain as (Column 2)

$$\varepsilon = \frac{\Delta L}{L_c} = \frac{\text{Column } 1}{\text{Line 10, Table } 18-3} \tag{18.13}$$

where ΔL = axial deformation

6. Calculate the piston load, P (Column 4)

$$P = (\text{proving ring dial reading, i.e. Column 3}) \times (\text{calibration factor}) \tag{18.14}$$

Table 18–3. Consolidated-Undrained Triaxial Test
Preliminary Data

Description of soil *Remolded grundite* Specimen No. *2*

Location

Tested by Date

Beginning of Test	
1. Moist unit weight of specimen (beginning of test)	
2. Moisture content (beginning of test)	*35.35%*
3. Initial length of specimen, L_0	*7.62 cm*
4. Initial diameter of specimen, D_0	*3.57 cm*
5. Initial area of the specimen, $A_0 = \frac{\pi}{4} D_0^2$	*10.0 cm²*
6. Initial volume of the specimen, $V_0 = A_0 L_0$	*76.2 cm³*
After Consolidation of Saturated Specimen	
7. Cell consolidation pressure, σ_3	*392 kN/m²*
8. Net drainage from the specimen during consolidation, ΔV	*11.6 cm³*
9. Volume of specimen after consolidation, $V_0 - \Delta V = V_c$	$76.2 - 11.6 = 64.6\ cm^3$
10. Length of the specimen after consolidation, $L_c = L_0\left(\frac{V_c}{V_0}\right)^{1/3}$	$7.62\left(\frac{64.6}{76.2}\right)^{1/3} = 7.212\ cm$
11. Area of the specimen after consolidation, $A_c = A_0\left(\frac{V_c}{V_0}\right)^{2/3}$	$10\left(\frac{64.6}{76.2}\right)^{2/3} = 8.96\ cm^3$

Table 18–4. Consolidated-Undrained Triaxial Test
Axial Stress-Strain Calculation

Proving ring calibration factor *1.0713 N/div.* (the results have been edited)

Specimen deformation $= \Delta L$ (cm) (1)	Vertical strain. $\varepsilon = \frac{\Delta L}{L_c}$ (2)	Proving ring dial reading (No. of small divisions) (3)	Piston load, P (N) (4)	Corrected area, $A = \frac{A_c}{1-\varepsilon}$ (cm^2) (5)	Deviatory stress, $\Delta\sigma = \frac{P}{A}$ (kN/m^2) (6)	Excess pore water pressure, Δu (kN/m^2) (7)	$\bar{A} = \frac{\Delta L}{\Delta\sigma}$ (8)
0	0	0	0	8.96	0	0	0
0.015	0.0021	15	16.07	8.98	17.90	2.94	0.164
0.038	0.0053	109	116.77	9.01	129.60	49.50	0.378
0.061	0.0085	135	144.63	9.04	159.99	74.56	0.466
0.076	0.0105	147	157.48	9.06	173.82	89.27	0.514
0.114	0.0158	172	184.26	9.11	202.26	111.83	0.553
0.152	0.0211	192	205.69	9.15	224.80	135.38	0.602
0.183	0.0254	205	219.62	9.19	238.98	148.13	0.620
0.229	0.0318	225	241.04	9.25	260.58	160.88	0.618
0.274	0.0380	236	252.83	9.31	271.57	167.75	0.618
0.315	0.0437	247	264.61	9.37	282.40	175.60	0.622
0.427	0.0592	265	283.89	9.52	298.20	176.58	0.592
0.457	0.0634	270	289.25	9.57	302.25	176.58	0.584
0.503	0.0697	278	297.82	9.63	309.26	176.58	0.570
0.549	0.0761	284	304.25	9.70	313.66	176.58	0.563
0.594	0.0824	287	307.46	9.76	315.02	176.58	0.561
0.653	0.905	288	308.53	9.85	313.23	176.57	0.564
0.726	0.1007	286	306.39	9.96	307.62	160.88	0.523
0.853	0.183	275	294.61	10.16	289.97	163.82	0.565

7. Calculate the corrected area, A, as (Column 5)

$$A = \frac{A_c}{1-\varepsilon} = \frac{\text{Line 11, Table 18-3}}{1 - \text{Column 2}} \tag{18.15}$$

8. Determine the deviatory stress, $\Delta\sigma$ (Column 6)

$$\Delta\sigma = \frac{P}{A} = \frac{\text{Column 4}}{\text{Column 5}} \tag{18.16}$$

9. Determine the pore water pressure parameter, $\overline{A}$ (Column 8)

$$\overline{A} = \frac{\Delta u}{\Delta\sigma} = \frac{\text{Column 7}}{\text{Column 6}} \tag{18.17}$$

10. Plot graphs for
    (a) $\Delta\sigma$ vs. ϵ (%)
    (b) Δu vs. ϵ (%)
    (c) $\overline{A}$ vs. ϵ (%)

As an example, the results of the calculation shown in Table 18–4 are plotted in Fig. 18–6.

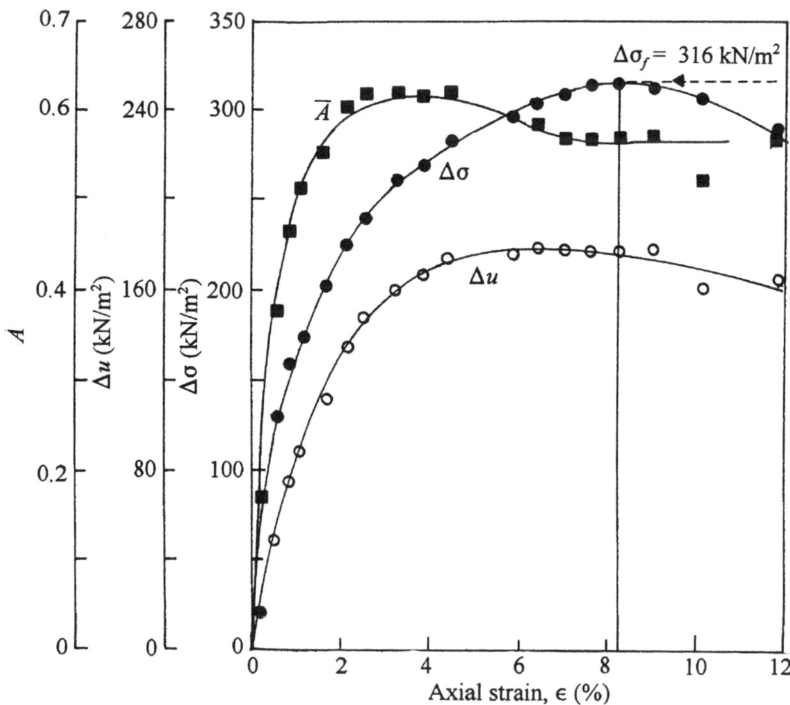

Figure 18–6. Plot of $\Delta\sigma$, Δu and $\overline{A}$ against axial strain for the consolidated drained test reported in Table 18–4.

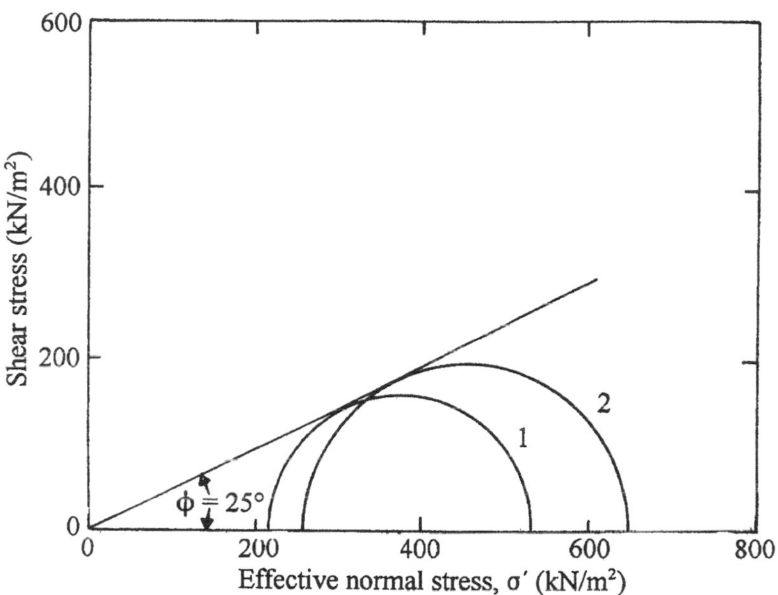

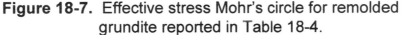

Figure 18-7. Effective stress Mohr's circle for remolded grundite reported in Table 18-4.

11. From the $\Delta\sigma$ vs. ϵ (%) graph, determine the maximum value of $\Delta\sigma = \Delta\sigma_f$ and the corresponding values of $\Delta u = \Delta u_f$ and $\overline{A} = \overline{A}_f$.

    In Fig. 18–6, $\Delta\sigma_f = 316$ kN/m^2 at $\epsilon = 8.2\%$ and, at the same strain level, $\Delta u_f = 177$ kN/m^2 and $\overline{A} = 0.56$.

12. Calculate the *effective* major and minor principal stresses at failure.

    Effective minor principal stress at failure

$$\sigma_3' = \sigma_3 - \Delta u_f \tag{18.18}$$

Effective major principal stress at failure

$$\sigma_1' = (\sigma_3 + \Delta\sigma_f) - \Delta u_f \tag{18.19}$$

For the test on the remolded grundite reported in Tables 18–3 and 18–4

$$\sigma_3' = 392 - 177 = 215 \text{ kN/m}^2$$

$$\sigma_1' = (392 + 316) - 177 = 531 \text{ kN/m}^2$$

13. Collect σ_1' and σ_3' for all the specimens tested and plot Mohr's circles. Plot a failure envelope that touches the Mohr's circles. The equation for the failure envelope can be given by

$$s = c + \sigma' \tan \phi$$

Determine the values of c and ϕ from the failure envelopes.
Figure 18–7 shows the Mohr's circles for two tests on the remolded grundite reported in Table 18–4. (*Note:* The result for the Mohr's circle No. 2 is not given in Table 18–4.) For the failure envelope, $c = 0$ and $\phi = 25°$. So

$$s = \sigma' \tan 25°$$

General Comments

1. For normally consolidated soils, $c = 0$; however, for overconsolidated soils, $c > 0$.
2. A typical range of values of $\overline{A}$ at failure for clayey soils is given below:

Type of Soil	$\overline{A}$ at Failure
Clays with high sensitivity	0.75 to 1.5
Normally consolidated clays	0.5 to 1.0
Overconsolidated clays	−0.5 to 0
Compacted sandy clay	0.5 to 0.75

REFERENCES

1. American Society for Testing and Materials, *1995 Annual Book of ASTM Standards* –Vol. 04.08, Philadelphia, PA, 1995.
2. Atterberg, A., "Über die Physikalishce Budenuntersuchung, and über die Plastizität der Töne," *Internationale Mitteilungen für Bodenkunde*, Vol. 1, 1911.
3. Casagrande, A., "Determination of Preconsolidation Load and Its Practical Significance," *Proceedings*, First International Conference on Soil Mechanics and Foundation Engineering, Vol. 3, 1963, pp. 60–64.
4. Casagrande, A. and Fadum, R.E., "Notes on Soil Testing for Engineering Purposes," *Engineering Publication No. 8*, Harvard University Graduate School, 1940.
5. Das, B.M., *Principles of Geotechnical Engineering, 3rd Edition*, PWS Publishing Company, Boston, 1994.
6. Proctor, R.R., "Design and Constsruction of Rolled Earth Dams," *Engineering News-Reconrd*, August 31, September 7, September 21, and September 28, 1933.
7. Rendon-Herrero, O., "University Compression Index Equation," *Journal of the Geotechnical Engineering Division*, American Society of Civil Engineers, Vol. 106, No. GT11, 1980, pp. 1179–1200.
8. Taylor, D.W., "Research on the Consolidation of Clays," *Serial No. 82*, Department of Civil and Sanitary Engineering, Massachusetts Institute of Technology, 1942.
9. Waterways Experiment Station, "Simplification of the Liquid Limit Test Procedure," *Technical Memorandum No. 3–286*, 1949.

APPENDIX A

Weight-Volume Relationships

For the weight-volume relationships given below, the following notions were used.

e = void ratio
G_s = specific gravity of soil solids
n = porosity
S = degree of saturation
V = total volume of soil
V_s = volume of solids in a soil mass
V_v = volume of voids in a soil mass
V_w = volume of water in a soil mass
W = total weight of a soil mass
W_s = dry weight of a soil mass
W_w = weight of water in a soil mass
w = moisture content
γ = moist unit weight
γ_d = dry unit weight
γ_{sat} = saturated unit weight
γ_w = unit weight of water

Volume Relationships

$$e = \frac{V_v}{V_s} = \frac{n}{1-n} = \frac{G_s \gamma_w}{\gamma_d} - 1$$

$$n = \frac{V_v}{V} = \frac{e}{1+e} = 1 - \frac{\gamma_d}{G_s \gamma_w}$$

$$S = \frac{V_w}{V_v} = \frac{w G_s}{e}$$

Weight Relationships

$$w = \frac{W_w}{W_s} = \frac{Se}{G_s}$$

$$\gamma = \frac{W}{V}$$

$$\gamma = \frac{G_s \gamma_w (1+w)}{1+e}$$

$$\gamma = G_s \gamma_w (1-n)(1+w)$$

$$\gamma_d = \frac{W_s}{V}$$

$$\gamma_d = \frac{G_s \gamma_w}{1+e}$$

$$\gamma_d = (1-n) G_s \gamma_w$$

$$\gamma_{\text{sat}} = \frac{(G_s + e)\gamma_w}{1+e}$$

$$\gamma_{\text{sat}} = \frac{(G_s + wG_s)\gamma_w}{1+wG_s} = \frac{G_s(1+w)\gamma_w}{1+wG_s}$$

$$\gamma_{\text{sat}} = [G_s - n(G_s - 1)]\gamma_w$$

APPENDIX B

Data Sheets for Laboratory Experiments

Determination of Water Content

Description of soil ____________________ Sample No. __________

Location ____________________

Tested by ____________________ Date __________

Item	Test No.		
	1	2	3
Can No.			
Mass of can, W_1 (g)			
Mass of can + wet soil, W_2 (g)			
Mass of can + dry soil, W_3 (g)			
Mass of moisture, $W_2 - W_3$ (g)			
Mass of dry soil, $W_3 - W_1$ (g)			
Moisture content, w (%) = $\dfrac{W_2 - W_3}{W_3 - W_1} \times 100$			

Average moisture content, w __________ %

Specific Gravity of Soil Solids

Description of soil____________________ Specimen No.________

Volume of flask at 20°C ______ ml Temperature of test __________°C A__________ (Table 3–2)

Location____________________

Tested by ____________________ Date____________________

Item	Test No.		
	1	2	3
Volumetric flask No.			
Mass of flask + water filled to mark, W_1 (g)			
Mass of flask + soil + water filled to mark, W_2 (g)			
Mass of dry soil, W_s (g)			
Mass of equal volume of water as the soil solids, W_w (g) = $(W_1 + W_s) - W_2$			
$G_{s(T_1 °C)} = W_s / W_w$			
$G_{s(20°C)} = G_{s(T_1 °C)} \times A$			

Average G_s ____________________

Sieve Analysis

Description of soil ____________________ Sample No. ________

Mass of oven dry sample, W ________ g

Location ____________________

Tested by ____________________ Date ____________

Sieve No.	Sieve opening (mm)	Mass of soil retained on each sieve, W_n (g)	Percent of mass retained on each sieve, R_n	Cumulative percent retained, $\sum R_n$	Percent finer, $100-\sum R_n$
Pan	—				

$\sum$ ________ $= W_1$

Mass loss during sieve analysis = $\dfrac{W - W_1}{W} \times 100$ = ________ % (OK if less than 2%)

Hydrometer Analysis

Description of soil ______________________________ Sample No.__________

Location ____

G_s ____________________ Hydrometer type ______________________

Dry weight of soil, W_s ________________ g Temperature of test, T ________________ °C

Meniscus correction, F_m ______ Zero correction, F_z ______ Temperature correction, F_T ________ [Eq. (5.6)]

Tested by______________________________ Date ____________________

Time (min) (1)	Hydrometer reading, R (2)	R_{cp} (3)	Percent finer, $\frac{a^{*} R_{cp}}{50} \times 100$ (4)	R_{cL} (5)	$L^{\dagger}$ (cm) (6)	$A^{\ddagger}$ (7)	D (mm) (8)

* Table 5.3; † Table 5.1; ‡ Table 5.2

Liquid Limit Test

Description of soil ______________________________ Sample No. __________

Location ______________________________

Tested by ______________________________ Date ______________

Test No.	1	2	3
Can No.			
Mass of can, W_1 (g)			
Mass of can + moist soil, W_2 (g)			
Mass of can + dry soil, W_3 (g)			
Moisture content, w (%) $= \dfrac{W_2 - W_3}{W_3 - W_1} \times 100$			
Number of blows, N			

Liquid limit = ______________________________

Flow index = ______________________________

Plastic Limit Test

Description of soil ______________________________ Sample No. __________

Location ____

Tested by ______________________________ Date ______________

Test No.	1	2	3
Can No.	103		
Mass of can, W_1 (g)			
Mass of can + moist soil, W_2 (g)			
Mass of can + dry soil, W_3 (g)			
$PL = \frac{W_2 - W_3}{W_3 - W_1} \times 100$			

Plasticity index, $PI = LL - PL =$ ______________ = ____________

Shrinkage Limit Test

Description of soil ______________________________ Sample No. ________

Location ____

Tested by ____________________________ Date ______________

Test No.		
Mass of coated shrinkage limit dish, W_1 (g)		
Mass of dish + wet soil, W_2 (g)		
Mass of dish + dry soil, W_3 (g)		
$w_i\ (\%) = \dfrac{(W_2 - W_3)}{(W_3 - W_1)} \times 100$		
Mass of mercury to fill the dish, W_4 (g)		
Mass of mercury displaced by soil pat, W_5 (g)		
$\Delta w_i\ (\%) = \dfrac{(W_4 - W_5)}{(13.6)(W_3 - W_1)} \times 100$		
$SL = w_i - \dfrac{(W_4 - W_5)}{(13.6)(W_3 - W_1)} \times 100$		

Constant Head Permeability Test

Determination of Void Ratio of Specimen

Description of soil ______________________________ Sample No. __________

Location ____

Length of specimen, L ______________cm Diameter of specimen, D __________cm

Tested by __________________________ Date______________________

Volume of specimen, $V = \frac{\pi}{4} D^2 L$ (cm^2)	
Specific gravity of soil solids, G_s	
Mass of specimen tube with fittings, W_1 (g)	
Mass of tube with fittings and specimen, W_2 (g)	
Dry density of specimen, $\rho_d = \frac{W_2 - W_1}{V}$ (g / cm^3)	
Void ratio of specimen, $e = \frac{G_s \rho_w}{\rho_d} - 1$ (*Note:* ρ_w = 1 g/cm^3)	

Constant Head Permeability Test

Determination of Coefficient of Permeability

Test No.	1	2	3
Average flow, Q (cm^3)			
Time of collection, t (s)			
Temperature of water, T (°C)			
Head difference, h (cm)			
Diameter of specimen, D (cm)			
Length of specimen, L (cm)			
Area of specimen, $A = \frac{\pi}{4} D^2$ (cm^2)			
$k = \frac{QL}{Aht}$ (cm / s)			
Average k = __________ cm/s			
$k_{20°C} = k_{T°C} \frac{\eta_{T°C}}{\eta_{20°C}}$ = ________________ = __________ cm/s			

Falling Head Permeability Test
Determination of Void Ratio of Specimen

Description of soil ______________________ Sample No. ____________

Location ____

Length of specimen, L ______________ cm Diameter of specimen, D __________ cm

Tested by ______________________ Date ____________

Volume of specimen, $V = \frac{\pi}{4} D^2 L \, (\text{cm}^2)$	
Specific gravity of soil solids, G_s	
Mass of specimen tube with fittings, W_1 (g)	
Mass of tube with fittings and specimen, W_2 (g)	
Dry density of specimen, $\rho_d = \frac{W_2 - W_1}{V} \, (\text{g/cm}^3)$	
Void ratio of specimen, $e = \frac{G_s \rho_w}{\rho_d} - 1$ (*Note:* $\rho_w = 1$ g/cm^3)	

Falling Head Permeability Test

Determination of Coefficient of Permeability

Test No.	1	2	3
Diameter of specimen, D (cm)			
Length of specimen, L (cm)			
Area of specimen, A (cm^2)			
Beginning head difference, h_1 (cm)			
Ending head difference, h_2 (cm)			
Test duration, t (s)			
Volume of water flow through the specimen, V_w (cm^3)			
$k = \dfrac{2.303 V_w L}{(h_1 - h_2) t A} \log \dfrac{h_1}{h_2}$ (cm / s^2)			
Average k = __________ cm/s			
$k_{20°C} = k_{T°C} \dfrac{\eta_{T°C}}{\eta_{20°C}}$ = ______________ = ________ cm/s			

Standard Proctor Compaction Test
Determination of Dry Unit Weight

Description of soil ______________________ Sample No. ________

Location ______________________________________

Volume of mold ________ ft^3 Weight of hammer ________ lb Number of blows/ layer ________ Number of layers ________

Tested by ______________________ Date ______________

Test	1	2	3	4	5	6
1. Weight of mold, W_1 (lb)						
2. Weight of mold + moist soil, W_2 (lb)						
3. Weight of moist soil, $W_2 - W_1$ (lb)						
4. Moist unit weight, $\gamma = \frac{W_2 - W_1}{1/30}$ (lb / ft^3)						
5. Moisture can number						
6. Mass of moisture can, W_3 (g)						
7. Mass of can + moist soil, W_4 (g)						
8. Mass of can + dry soil, W_5 (g)						
9. Moisture content, $w\ (\%) = \frac{W_4 - W_5}{W_5 - W_3} \times 100$						
10. Dry unit weight of compaction $\gamma_d\ (\text{lb}/\text{ft}^3) = \frac{\gamma}{1 + \frac{w\ (\%)}{100}}$						

Standard Proctor Compaction Test

Zero-Air-Void Unit Weight

Description of soil ________________________________ Sample No. ____________

Location ____

Tested by ______________________________ Date ______________

Specific gravity of soil solids, G_s	Assumed moisture content, w (%)	Unit weight of water, γ_w (lb/ft^3)	γ_{zav} [a] (lb/ft^3)

[a] Eq. (12.1)

Modified Proctor Compaction Test

Determination of Dry Unit Weight

Description of soil ______________________ Sample No. __________

Location ______________________

Volume of mold __________ ft^3 Weight of hammer ________ lb Number of blows/ layer ________ Number of layers ________

Tested by ______________________ Date __________

Test	1	2	3	4	5	6
1. Weight of mold, W_1 (lb)						
2. Weight of mold + moist soil, W_2 (lb)						
3. Weight of moist soil, $W_2 - W_1$ (lb)						
4. Moist unit weight, $\gamma = \frac{W_2 - W_1}{1/30}$ (lb / ft^3)						
5. Moisture can number						
6. Mass of moisture can, W_3 (g)						
7. Mass of can + moist soil, W_4 (g)						
8. Mass of can + dry soil, W_5 (g)						
9. Moisture content, $w\ (\%) = \frac{W_4 - W_5}{W_5 - W_3} \times 100$						
10. Dry unit weight of compaction $\gamma_d\ (\text{lb}/\text{ft}^3) = \frac{\gamma}{1 + \frac{w\ (\%)}{100}}$						

Modified Proctor Compaction Test

Zero-Air-Void Unit Weight

Description of soil ________________________________ Sample No. __________

Location ____

Tested by ______________________________ Date ____________________

Specific gravity of soil solids, G_s	Assumed moisture content, w (%)	Unit weight of water, γ_w (lb/ft^3)	γ_{zav}[a] (lb/ft^3)

[a] Eq. (12.1)

Field Unit Weight – Sand Cone Method

Item	Quantity
Calibration of Unit Weight of Ottawa Sand	
1. Weight of Proctor mold, W_1	
2. Weight of Proctor mold + sand, W_2	
3. Volume of mold, V_1	
4. Dry unit weight, $\gamma_{d(\text{sand})} = \dfrac{W_2 - W_1}{V_1}$	
Calibration Cone	
5. Weight of bottle + cone + sand (before use), W_3	
6. Weight of bottle + cone + sand (after use), W_4	
7. Weight of sand to fill the cone, $W_c = W_4 - W_3$	
Results from Field Tests	
8. Weight of bottle + cone + sand (before use), W_6	
9. Weight of bottle + cone + sand (after use), W_8	
10 Volume of hole, $V_2 = \dfrac{W_6 - W_8 - W_c}{\gamma_{d(\text{sand})}}$.	
11 Weight of gallon can, W_5	
12. Weight of gallon can + moist soil, W_7	
13. Weight of gallon can + dry soil, W_9	
14. Moist unit weight of soil in field, $\gamma = \dfrac{W_7 - W_5}{V_2}$	
15. Moisture content in the field, $w\,(\%) = \dfrac{W_7 - W_9}{W_9 - W_5} \times 100$	
16. Dry unit weight in the field, $\gamma_{d(sand)} = \dfrac{\gamma}{1 + \dfrac{w\,(\%)}{100}}$	

Direct Shear Test on Sand
Void Shear Calculation

Description of soil ______________________ Sample No. __________

Location ______________________

Tested by ______________________ Date __________

Item	Quantity
1. Specimen length, L (in.)	
2. Specimen width, B (in.)	
3. Specimen height, H (in.)	
4. Mass of porcelain dish + dry sand (before use), W_1 (g)	
5. Mass of porcelain dish + dry sand (after use), W_2 (g)	
6. Dry unit weight of specimen, $\gamma_d\ (\text{lb}/\text{ft}^3) = \dfrac{W_1 - W_2\ (\text{g})}{LBH\,(\text{in.}^3)} \times 3.808$	
7. Specific gravity of soil solids, G_s	
8. Void ratio, $e = \dfrac{G_s \gamma_w}{\gamma_d} - 1$ *Note:* $\gamma_w = 62.4\ \text{lb/ft}^3$; γ_d is in lb/ft^3	

Direct Shear Test on Sand

Stress and Displacement Calculation

Description of soil ______________________________ Sample No. ____________

Location ____

Normal load, N ____________________ lb Void ratio, e ____________________

Tested by ______________________________ Date ____________________

Normal stress, σ' (lb/in.^{2}) (1)	Horizontal displacement (in.) (2)	Vertical displacement (in.) (3)	No. of div. in proving ring dial gauge (4)	Proving ring calibration factor (lb/div) (5)	Shear force, S (lb) (6)	Shear stress, τ (lb/in.^{2}) (7)

* Plus (+) sign means expansion

Direct Shear Test on Sand

Stress and Displacement Calculation

Description of soil ________________________ Sample No. __________

Location ____

Normal load, N ______________ lb Void ratio, e ______________

Tested by ______________________ Date ______________

Normal stress, σ' (lb/in.^{2}) (1)	Horizontal displacement (in.) (2)	Vertical displacement (in.) (3)	No. of div. in proving ring dial gauge (4)	Proving ring calibration factor (lb/div) (5)	Shear force, S (lb) (6)	Shear stress, τ (lb/in.^{2}) (7)

* Plus (+) sign means expansion

Direct Shear Test on Sand

Stress and Displacement Calculation

Description of soil ______________________ Sample No. __________

Location ______________________________________

Normal load, N ______________ lb Void ratio, e ______________

Tested by ______________________ Date ______________

Normal stress, σ' (lb/in.^{2}) (1)	Horizontal displacement (in.) (2)	Vertical displacement (in.) (3)	No. of div. in proving ring dial gauge (4)	Proving ring calibration factor (lb/div) (5)	Shear force, S (lb) (6)	Shear stress, τ (lb/in.^{2}) (7)

* Plus (+) sign means expansion

Unconfined Compression Test

Description of soil ______________________________ Specimen No. __________

Location ____

Moist weight of specimen ________ g Moisture content _____ % Length of specimen, L ______ in. Diameter of specimen ________ in.

Proving ring calibration factor: 1 div. = __________ lb Area, $A_0 = \frac{\pi}{4} D^2$ = ____________ in.^{2}

Tested by ______________________________ Date ______________

Specimen deformation $= \Delta L$ (in.) (1)	Vertical strain, $\epsilon = \frac{\Delta L}{L}$ (2)	Proving ring dial reading [No. of small divisions] (3)	Load = Column 3 × calibration factor of proving ring (lb) (4)	Corrected area, $A_c = \frac{A_0}{1 - \epsilon}$ (in.^{2}) (5)	Stress, $\sigma = \frac{\text{Column 4}}{\text{Column 5}}$ (lb/in.^{2}) (6)

Consolidation Test
Time vs. Vertical Dial Reading

Description of soil ______________________________

Location ______________________________

Tested by ______________________________ Date ______________

Pressure on specimen _____ lb/ft^2 Pressure on specimen _____ lb/ft^2

Time after load application, t (min.)	$\sqrt{t}$ (min.)$^{0.5}$	Vertical dial reading (in.)	Time after load application, t (min.)	$\sqrt{t}$ (min.)$^{0.5}$	Vertical dial reading (in.)

Consolidation Test

Time vs. Vertical Dial Reading

Description of soil ______________________________

Location ______________________________

Tested by ______________________ Date ______________

Pressure on specimen _____ lb/ft^2			Pressure on specimen _____ lb/ft^2		
Time after load application, t (min.)	$\sqrt{t}$ (min.)$^{0.5}$	Vertical dial reading (in.)	Time after load application, t (min.)	$\sqrt{t}$ (min.)$^{0.5}$	Vertical dial reading (in.)

Consolidation Test

Time vs. Vertical Dial Reading

Description of soil ______________________________

Location ______________________________

Tested by ______________________________ Date ______________

Pressure on specimen _____ lb/ft^2			Pressure on specimen _____ lb/ft^2		
Time after load application, t (min.)	$\sqrt{t}$ (min.)$^{0.5}$	Vertical dial reading (in.)	Time after load application, t (min.)	$\sqrt{t}$ (min.)$^{0.5}$	Vertical dial reading (in.)

Consolidation Test

Void Ratio-Pressure and Coefficient of Consolidation Calculation

Description of soil ______________________________ Location ______________________________

Specimen diameter _____ in. Initial specimen height, $H_{t(i)}$ _____ in. Height of solids, H_s ____ cm = ____ in.

Moisture Content: Beginning of test _______ % End of test _________ %

Weight of dry soil specimen ____________ g G_s ___________

Tested by __________________________________ Date ______________________________________

Pressure p (ton/ft^2) (1)	Final dial reading (in.) (2)	Change in specimen height, ΔH (in.) (3)	Final specimen height, $H_{t(f)}$ (in.) (4)	Height of void, H_v (in.) (5)	Final void ratio, e (6)	Average height during consolidation, $H_{t(av)}$ (in.) (7)	Fitting time (sec)		c_v from × 10^3 (in.^{2}/sec)	
							t_{90} (8)	t_{50} (9)	t_{90} (10)	t_{50} (11)

Unconsolidated-Undrained Triaxial Test

Preliminary Data

Description of soil ______________________ Specimen No. __________

Location ____

Tested by ______________________ Date __________

Item	Quantity
1. Moist mass of specimen (end of test), W_1	
2. Dry mass of specimen, W_2	
3. Moisture content (end of test), $w\ (\%) = \dfrac{W_1 - W_2}{W_2} \times 100$	
4. Initial average length of specimen, L_0	
5. Initial average diameter of specimen, D_0	
6. Initial area, $A_0 = \dfrac{\pi}{4} D^2$	
7. Specific gravity of soil solids, G_s	
8. Final degree of saturation	
9. Cell confining pressure, σ_3	
10. Proving ring calibration factor	

Unconsolidated-Undrained Triaxial Test

Axial Stress-Strain Calculation

Specimen deformation $= \Delta L$ (in.) (1)	Vertical strain, $= \frac{\Delta L}{L_0}$ (2)	Proving ring dial reading (Number of small divisions) (3)	Piston load, P (Column 3 × calibration factor) lb (4)	Corrected area, $A = \frac{A_0}{1 - \epsilon}$ (in.^{2}) (5)	Deviatory stress, $\Delta\sigma = \frac{P}{A}$ (lb/in.^{2}) (6)

Consolidated-Undrained Triaxial Test

Preliminary Data

Description of soil ____________________ Specimen No. ____________

Location ____

Tested by ____________________ Date ____________

Beginning of Test	
1. Moist unit weight of specimen (beginning of test)	
2. Moisture content (beginning of test)	
3. Initial length of specimen, L_0	
4. Initial diameter of specimen, D_0	
5. Initial area of the specimen, $A_0 = \frac{\pi}{4} D_0^2$	
6. Initial volume of the specimen, $V_0 = A_0 L_0$	
After Consolidation of Saturated Specimen	
7. Cell consolidation pressure, σ_3	
8. Net drainage from the specimen during consolidation, ΔV	
9. Volume of specimen after consolidation, $V_0 - \Delta V = V_c$	
10. Length of the specimen after consolidation, $L_c = L_0 \left(\frac{V_c}{V_0} \right)^{1/3}$	
11. Area of the specimen after consolidation, $A_c = A_0 \left(\frac{V_c}{V_0} \right)^{2/3}$	

Consolidated-Undrained Triaxial Test

Axial Stress-Strain Calculation

Proving ring calibration factor ____________________

Specimen deformation $= \Delta L$ (cm) (1)	Vertical strain, $\epsilon = \frac{\Delta L}{L_c}$ (2)	Proving ring dial reading (No. of small divisions) (3)	Piston load, P (N) (4)	Corrected area, $A = \frac{A_c}{1-\epsilon}$ (cm^2) (5)	Deviatory stress, $\Delta\sigma = \frac{P}{A}$ (6)	Excess pore water pressure, Δu (kN/m^2) (7)	$\bar{A} = \frac{\Delta L}{\Delta\sigma}$ (8)

Consolidated-Undrained Triaxial Test

Preliminary Data

Description of soil ________________________________ Specimen No. ____________

Location ____

Tested by ____ Date ____________

Beginning of Test	
1. Moist unit weight of specimen (beginning of test)	
2. Moisture content (beginning of test)	
3. Initial length of specimen, L_0	
4. Initial diameter of specimen, D_0	
5. Initial area of the specimen, $A_0 = \frac{\pi}{4} D_0^2$	
6. Initial volume of the specimen, $V_0 = A_0 L_0$	
After Consolidation of Saturated Specimen	
7. Cell consolidation pressure, σ_3	
8. Net drainage from the specimen during consolidation, ΔV	
9. Volume of specimen after consolidation, $V_0 - \Delta V = V_c$	
10. Length of the specimen after consolidation, $L_c = L_0 \left(\frac{V_c}{V_0} \right)^{1/3}$	
11. Area of the specimen after consolidation, $A_c = A_0 \left(\frac{V_c}{V_0} \right)^{2/3}$	

Consolidated-Undrained Triaxial Test
Axial Stress-Strain Calculation

Proving ring calibration factor ______________________

Specimen deformation $= \Delta L$ (cm) (1)	Vertical strain. $\epsilon = \frac{\Delta L}{L_c}$ (2)	Proving ring dial reading (No. of small divisions) (3)	Piston load, P (N) (4)	Corrected area, $A = \frac{A_c}{1-\epsilon}$ (cm^2) (5)	Deviatory stress, $\Delta\sigma = \frac{P}{A}$ (6)	Excess pore water pressure, Δu (kN/m^2) (7)	$\overline{A} = \frac{\Delta L}{\Delta\sigma}$ (8)

APPENDIX C

Data Sheets for Preparation of Laboratory Reports

Specific Gravity of Soil Solids

Description of soil____________________ Specimen No.________

Volume of flask at 20°C ______ ml Temperature of test __________°C A__________ (Table 3–2)

Location____________________

Tested by ____________________ Date____________

Item	Test No.		
	1	2	3
Volumetric flask No.			
Mass of flask + water filled to mark, W_1 (g)			
Mass of flask + soil + water filled to mark, W_2 (g)			
Mass of dry soil, W_s (g)			
Mass of equal volume of water as thc soil solids, W_w (g) = $(W_1 + W_s) - W_2$			
$G_{s(T_1°C)} = W_s / W_w$			
$G_{s(20°C)} = G_{s(T_1°C)} \times A$			

Average G_s ____________

Shrinkage Limit Test

Description of soil ______________________ Sample No. ________

Location ______________________

Tested by ______________________ Date ________

Test No.		
Mass of coated shrinkage limit dish, W_1 (g)		
Mass of dish + wet soil, W_2 (g)		
Mass of dish + dry soil, W_3 (g)		
$w_i\ (\%) = \dfrac{(W_2 - W_3)}{(W_3 - W_1)} \times 100$		
Mass of mercury to fill the dish, W_4 (g)		
Mass of mercury displaced by soil pat, W_5 (g)		
$\Delta w_i\ (\%) = \dfrac{(W_4 - W_5)}{(13.6)(W_3 - W_1)} \times 100$		
$SL = w_i - \dfrac{(W_4 - W_5)}{(13.6)(W_3 - W_1)} \times 100$		

Falling Head Permeability Test

Determination of Void Ratio of Specimen

Description of soil ______________________ Sample No. ______________

Location ____

Length of specimen, L ______________ cm Diameter of specimen, D ______________ cm

Tested by ______________________ Date ______________

Volume of specimen, $V = \frac{\pi}{4} D^2 L\,(\text{cm}^2)$	
Specific gravity of soil solids, G_s	
Mass of specimen tube with fittings, W_1 (g)	
Mass of tube with fittings and specimen, W_2 (g)	
Dry density of specimen, $\rho_d = \frac{W_2 - W_1}{V}\,(\text{g}/\text{cm}^3)$	
Void ratio of specimen, $e = \frac{G_s \rho_w}{\rho_d} - 1$ (*Note:* $\rho_w = 1$ g/cm^3)	

Falling Head Permeability Test

Determination of Coefficient of Permeability

Test No.	1	2	3
Diameter of specimen, D (cm)			
Length of specimen, L (cm)			
Area of specimen, A (cm^2)			
Beginning head difference, h_1 (cm)			
Ending head difference, h_2 (cm)			
Test duration, t (s)			
Volume of water flow through the specimen, V_w (cm^3)			
$k = \dfrac{2.303 V_w L}{(h_1 - h_2) t A} \log \dfrac{h_1}{h_2}$ (cm / s^2)			
Average k = ___________ cm/s			
$k_{20°C} = k_{T°C} \dfrac{\eta_{T°C}}{\eta_{20°C}}$ = ________________ = ________ cm/s			

Modified Proctor Compaction Test

Determination of Dry Unit Weight

Description of soil ______________________ Sample No. __________

Location ______________________

Volume of mold __________ ft^3 Weight of hammer ________ lb Number of blows/layer ________ Number of layers ________

Tested by ______________________ Date __________

Test	1	2	3	4	5	6
1. Weight of mold, W_1 (lb)						
2. Weight of mold + moist soil, W_2 (lb)						
3. Weight of moist soil, $W_2 - W_1$ (lb)						
4. Moist unit weight, $\gamma = \dfrac{W_2 - W_1}{1/30}$ (lb / ft^3)						
5. Moisture can number						
6. Mass of moisture can, W_3 (g)						
7. Mass of can + moist soil, W_4 (g)						
8. Mass of can + dry soil, W_5 (g)						
9. Moisture content, $w\,(\%) = \dfrac{W_4 - W_5}{W_5 - W_3} \times 100$						
10. Dry unit weight of compaction $\gamma_d\,(\text{lb/ft}^3) = \dfrac{\gamma}{1 + \dfrac{w\,(\%)}{100}}$						

Modified Proctor Compaction Test

Zero-Air-Void Unit Weight

Description of soil ______________________ Sample No. __________

Location ______________________

Tested by ______________________ Date __________

Specific gravity of soil solids, G_s	Assumed moisture content, w (%)	Unit weight of water, γ_w (lb/ft^3)	γ_{zav}[a] (lb/ft^3)

[a] Eq. (12.1)

Field Unit Weight – Sand Cone Method

Item	Quantity
Calibration of Unit Weight of Ottawa Sand	
1. Weight of Proctor mold, W_1	
2. Weight of Proctor mold + sand, W_2	
3. Volume of mold, V_1	
4. Dry unit weight, $\gamma_{d(\text{sand})} = \dfrac{W_2 - W_1}{V_1}$	
Calibration Cone	
5. Weight of bottle + cone + sand (before use), W_3	
6. Weight of bottle + cone + sand (after use), W_4	
7. Weight of sand to fill the cone, $W_c = W_4 - W_3$	
Results from Field Tests	
8. Weight of bottle + cone + sand (before use), W_6	
9. Weight of bottle + cone + sand (after use), W_8	
10 Volume of hole, $V_2 = \dfrac{W_6 - W_8 - W_c}{\gamma_{d(\text{sand})}}$.	
11. Weight of gallon can, W_5	
12. Weight of gallon can + moist soil, W_7	
13. Weight of gallon can + dry soil, W_9	
14. Moist unit weight of soil in field, $\gamma = \dfrac{W_7 - W_5}{V_2}$	
15. Moisture content in the field, $w(\%) = \dfrac{W_7 - W_9}{W_9 - W_5} \times 100$	
16. Dry unit weight in the field, $\gamma_{d(sand)} = \dfrac{\gamma}{1 + \dfrac{w(\%)}{100}}$	

Unconfined Compression Test

Description of soil ______________________ Specimen No. __________

Location ______________________________________

Moist weight of specimen __________ g Moisture content ______ % Length of specimen, L ______ in. Diameter of specimen ________ in.

Proving ring calibration factor: 1 div. = __________ lb Area, $A_0 = \frac{\pi}{4}D^2$ = ____________ in.^{2}

Tested by ______________________________ Date ______________

Specimen deformation $= \Delta L$ (in.) (1)	Vertical strain, $\epsilon = \frac{\Delta L}{L}$ (2)	Proving ring dial reading [No. of small divisions] (3)	Load = Column 3 × calibration factor of proving ring (lb) (4)	Corrected area, $A_c = \frac{A_0}{1-\epsilon}$ (in.^{2}) (5)	Stress, $\sigma = \frac{\text{Column 4}}{\text{Column 5}}$ (lb/in.^{2}) (6)

Consolidation Test

Time vs. Vertical Dial Reading

Description of soil ______________________________

Location ______________________________

Tested by ______________________________ Date ______________

Pressure on specimen ____ lb/ft^2 Pressure on specimen ____ lb/ft^2

Time after load application, t (min.)	$\sqrt{t}$ (min.)$^{0.5}$	Vertical dial reading (in.)	Time after load application, t (min.)	$\sqrt{t}$ (min.)$^{0.5}$	Vertical dial reading (in.)

Consolidation Test

Time vs. Vertical Dial Reading

Description of soil ______________________________

Location ______________________________

Tested by ______________________________ Date ______________

Pressure on specimen ____ lb/ft^2 | Pressure on specimen ____ lb/ft^2

Time after load application, t (min.)	$\sqrt{t}$ (min.)$^{0.5}$	Vertical dial reading (in.)	Time after load application, t (min.)	$\sqrt{t}$ (min.)$^{0.5}$	Vertical dial reading (in.)

Consolidation Test

Time vs. Vertical Dial Reading

Description of soil ______________________________

Location ______________________________

Tested by ______________________ Date ______________

Pressure on specimen ____ lb/ft²			Pressure on specimen ____ lb/ft²		
Time after load application, t (min.)	$\sqrt{t}$ (min.)$^{0.5}$	Vertical dial reading (in.)	Time after load application, t (min.)	$\sqrt{t}$ (min.)$^{0.5}$	Vertical dial reading (in.)

Consolidation Test

Void Ratio-Pressure and Coefficient of Consolidation Calculation

Description of soil ______________________________ Location ______________________________

Specimen diameter _____ in. Initial specimen height, $H_{t(i)}$ _____ in. Height of solids, H_s ___ cm = ____ in.

Moisture Content: Beginning of test _______ % End of test _________ %

Weight of dry soil specimen ___________ g G_s ___________

Tested by ______________________________ Date ______________________________

Pressure p (ton/ft^2) (1)	Final dial reading (in.) (2)	Change in specimen height, ΔH (in.) (3)	Final specimen height, $H_{t(f)}$ (in.) (4)	Height of void, H_v (in.) (5)	Final void ratio, e (6)	Average height during consolidation, $H_{t(av)}$ (in.) (7)	Fitting time (sec)		c_v from × 10^3 (in.^{2}/sec)	
							t_{90} (8)	t_{50} (9)	t_{90} (10)	t_{50} (11)

Unconsolidated-Undrained Triaxial Test

Preliminary Data

Description of soil ______________________ Specimen No. __________

Location ______________________________________

Tested by ______________________ Date __________

Item	Quantity
1. Moist mass of specimen (end of test), W_1	
2. Dry mass of specimen, W_2	
3. Moisture content (end of test), $w\,(\%) = \frac{W_1 - W_2}{W_2} \times 100$	
4. Initial average length of specimen, L_0	
5. Initial average diameter of specimen, D_0	
6. Initial area, $A_0 = \frac{\pi}{4} D^2$	
7. Specific gravity of soil solids, G_s	
8. Final degree of saturation	
9. Cell confining prcssure, σ_3	
10. Proving ring calibration factor	

Unconsolidated-Undrained Triaxial Test

Axial Stress-Strain Calculation

Specimen deformation $= \Delta L$ (in.) (1)	Vertical strain, $\epsilon = \frac{\Delta L}{L_n}$ (2)	Proving ring dial reading (Number of small divisions) (3)	Piston load, P (Column 3 × calibration factor) lb (4)	Corrected area, $A = \frac{A_0}{1-\epsilon}$ (in.^{2}) (5)	Deviatory stress, $\Delta\sigma = \frac{P}{A}$ (lb/in.^{2}) (6)

Constant Head Permeability Test

Determination of Void Ratio of Specimen

Description of soil Uniform Sand Sample No. ____________

Location ____________

Length of specimen, L 45 cm cm Diameter of specimen, D 29 cm cm

Tested by Weston Philips, Ryan Eisele, Scott Almeida, Brad Pleima Date 2/11/03

Volume of specimen, $V = \frac{\pi}{4} D^2 L$ (cm^2) = $\frac{\pi}{4}(15.3^2)($	
Specific gravity of soil solids, G_s	
Mass of specimen tube with fittings, W_1 (g)	
Mass of tube with fittings and specimen, W_2 (g)	
Dry density of specimen, $\rho_d = \frac{W_2 - W_1}{V}$ (g / cm^3)	
Void ratio of specimen, $e = \frac{G_s \rho_w}{\rho_d} - 1$ (*Note:* $\rho_w = 1$ g/cm^3)	

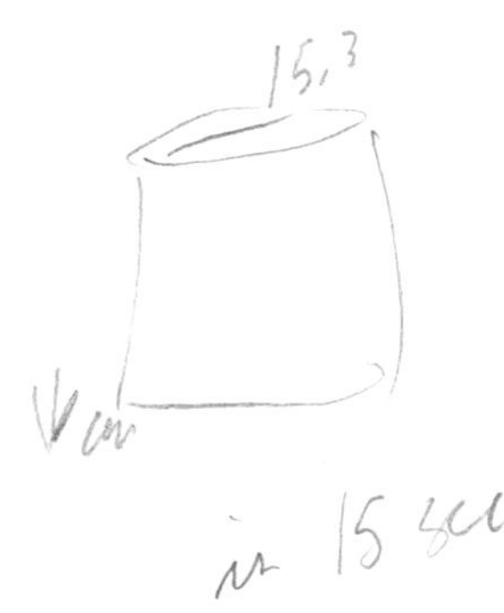

Consolidated-Undrained Triaxial Test

Preliminary Data

Description of soil ______________________ Specimen No. __________

Location ______________________________________

Tested by ______________________ Date __________

Beginning of Test	
1. Moist unit weight of specimen (beginning of test)	
2. Moisture content (beginning of test)	
3. Initial length of specimen, L_0	
4. Initial diameter of specimen, D_0	
5. Initial area of the specimen, $A_0 = \frac{\pi}{4} D_0^2$	
6. Initial volume of the specimen, $V_0 = A_0 L_0$	
After Consolidation of Saturated Specimen	
7. Cell consolidation pressure, σ_3	
8. Net drainage from the specimen during consolidation, ΔV	
9. Volume of specimen after consolidation, $V_0 - \Delta V = V_c$	
10. Length of the specimen after consolidation, $L_c = L_0 \left(\frac{V_c}{V_0} \right)^{1/3}$	
11. Area of the specimen after consolidation, $A_c = A_0 \left(\frac{V_c}{V_0} \right)^{2/3}$	

Consolidated-Undrained Triaxial Test

Axial Stress-Strain Calculation

Proving ring calibration factor ____________________

Specimen deformation $= \Delta L$ (1)	Vertical strain. $\epsilon = \frac{\Delta L}{L_c}$ (2)	Proving ring dial reading (No. of small divisions) (3)	Piston load, P (4)	Corrected area, $A = \frac{A_c}{1 - \epsilon}$ (5)	Deviatory stress, $\Delta\sigma = \frac{P}{A}$ (6)	Excess pore water pressure, Δu (7)	$\overline{A} = \frac{\Delta L}{\Delta\sigma}$ (8)

Consolidated-Undrained Triaxial Test

Preliminary Data

Description of soil ______________________ Specimen No. ____________

Location ____

Tested by ______________________ Date ____________

Beginning of Test	
1. Moist unit weight of specimen (beginning of test)	
2. Moisture content (beginning of test)	
3. Initial length of specimen, L_0	
4. Initial diameter of specimen, D_0	
5. Initial area of the specimen, $A_0 = \frac{\pi}{4} D_0^2$	
6. Initial volume of the specimen, $V_0 = A_0 L_0$	
After Consolidation of Saturated Specimen	
7. Cell consolidation pressure, σ_3	
8. Net drainage from the specimen during consolidation, ΔV	
9. Volume of specimen after consolidation, $V_0 - \Delta V = V_c$	
10. Length of the specimen after consolidation, $L_c = L_0 \left(\frac{V_c}{V_0} \right)^{1/3}$	
11. Area of the specimen after consolidation, $A_c = A_0 \left(\frac{V_c}{V_0} \right)^{2/3}$	

Consolidated-Undrained Triaxial Test
Axial Stress-Strain Calculation

Proving ring calibration factor ______________________

Specimen deformation $= \Delta L$ (1)	Vertical strain. $\epsilon = \frac{\Delta L}{L_c}$ (2)	Proving ring dial reading (No. of small divisions) (3)	Piston load, P (4)	Corrected area, $A = \frac{A_c}{1-\epsilon}$ (5)	Deviatory stress, $\Delta\sigma = \frac{P}{A}$ (6)	Excess pore water pressure, Δu (7)	$\overline{A} = \frac{\Delta L}{\Delta\sigma}$ (8)

SOFTWARE INSTALLATION

The Soil Mechanics Laboratory Test software for Windows requires an 80286 or better computer running Windows 3.1 or later, with at least 1 MB of free RAM. The installation process temporarily requires 4 MB of hard disk space, and permanently takes up 3 MB of hard disk space.

1. Windows installation is started using the Program Manager. Select the "File | Run" menu item ("Start | Run" on Windows 95 or later), and enter A:\SETUP (or B:\SETUP if you are installing from the B: floppy), then press the <Enter> key. Be sure to leave the diskette in the drive until step 8.

2. The Setup program will ask you for the directory into which you wish the files to be placed. The default directory is C:\SOILMECH, but you may select any name (only up to 8 characters long in Windows 3.1). The new directory will be automatically created. Once you select the directory name, the Setup program will copy and decompress all of the files to the hard disk, creating a program group with the name Soil Mechanics, including two entries within the group to access two programs, "File Translator" and "Laboratory Tests."

3. The Setup program will then start the File Translator program. Alternatively, you may at a later point resume the process by double clicking on the File Translator program icon from within the Program Manager. You must run this program because the files that have been copied are encoded; this program is needed to translate the files.

4. Select the "File | Register Yourself" menu entry. This will cause a dialog box to appear asking you to enter your name, address, and phone number.

5. Fill in ALL of the information fields. The program will not proceed until all fields have an entry. When you are ready, click on OK. The software will return to the main program screen, but the "File | Translate Files" menu entry will now be activated.

6. Select the "File | Translate Files" menu entry. The software will start translating and storing the program and data files, showing a message as each file is handled.

7. Select the "File | Exit" menu entry to quit the File Translator program. Your files have been translated so that your system can now work with them.

8. You can now remove the diskette from your drive and save it.

EXCEL SPREADSHEETS

The installation places 18 Excel worksheets into a subdirectory called "EXCEL." These can be used with Excel97 or Excel2000. (They will not appear properly under Excel95.) If Excel is already installed on your computer, you can double click on the icon for any of the worksheets to begin. When opening a file, make sure to Enable Macros, or the formulae will not function. Each of the worksheets starts with the values used in the textbook examples. You can change any of the fields that are in white—Excel will then update the values in the computed fields shown in yellow. You can show the plots by clicking on the button on the bottom of the worksheet that says "Show Plot." You can return to the data by clicking on the button on the graph that says "Show Data." You can print the visible display by clicking on the button that says "Print Test" or "Print Plot." Either of these buttons will take you to the Print Preview page, where you may review what is to print before pressing the "Print" button at the top of the preview page.

SOIL MECHANICS LABORATORY TEST SOFTWARE OPERATION

Start the Soil Mechanics Laboratory Test software by double clicking on its icon to bring up the main program window. Eighteen different soil mechanics tests are available to you by selecting one of the entries under the SOIL TESTS or MORE SOIL TESTS entry. Each test requires data to be entered, and displays the computed results. The results can be printed or saved to a file. The latest values for each of the tests is maintained while running the program, so that you can run each test simultaneously. You can save your analysis data by selecting the "File | Save" entry, then selecting the type of test data to be saved, and then selecting the filename. You must do a separate save operation for each type of test (saved to different files). Select the "File | Exit" entry to exit the program. Select the "Help | Guide to Soil Test" entry for additional operation information. This brings up the SOILHELP.TXT file; you can also just read/print this file with any word processor/editor.